Fundamentals of

ENVIRONMENTAL DISCHARGE MODELING

CRC MECHANICAL ENGINEERING SERIES

Frank Kreith, Series Editor
University of Colorado

Published Titles

Entropy Generation Minimization
  Adrian Bejan, Duke University
Nonlinear Analysis of Structures
  M. Sathyamoorthy, Clarkson University
Finite Element Method Using MATLAB
  Young W. Kwon, Naval Postgraduate School
  Hyochoong Bang, Korea Aerospace Research Institute
Mechanics of Composite Materials
  Autar K. Kaw, University of South Florida
Practical Inverse Analysis in Engineering
  David M. Trujillo, TRUCOMP
  Henry R. Busby, Ohio State University
Viscoelastic Solids
  Roderic S. Lakes

Forthcoming Titles

Mechanics of Fatigue
  Vladimir Bolotin
Engineering Experimentation
  Euan Somerscales
Mathematical and Physical Modeling of Materials Processing Operations
  Olusegun Johnson Ilegbusi, Northeastern University
  Manabu Iguchi, Osaka University
  Walter Wahnsiedler, ALOCOA Technical Center
Mechanics of Solids & Shells
  Gerals Wempner, Demosthenes Talaslidis
Energy Audit of Building Systems: An Engineering Approach
  Moncef Krarti

Fundamentals of

ENVIRONMENTAL DISCHARGE MODELING

Lorin R. Davis

Professor of Mechanical Engineering
Oregon State University
Corvallis

CRC Press
Taylor & Francis Group
Boca Raton London New York

CRC Press is an imprint of the
Taylor & Francis Group, an **informa** business

CRC Press
Taylor & Francis Group
6000 Broken Sound Parkway NW, Suite 300
Boca Raton, FL 3487-2742

First issued in paperback 2020

ISBN-13: 978-0-367-57924-1 (pbk)
ISBN-13: 978-0-8493-9657-1 (hbk)

Visit the Taylor & Francis Web site at
http://www.taylorandfrancis.com

and the CRC Press Web site at
http://www.crcpress.com

Library of Congress Cataloging-in-Publication Data

Catalog record is available from the Library of Congress.

Preface

The disposal of waste water in the environment is carefully controlled. Concentrations of pollutants are not allowed to exceed specified limits within prescribed mixing zones. In some cases no mixing zone is allowed and concentrations of all pollutants must be below set limits within the effluent. In those cases where a mixing zone is allowed, documentation must be supplied showing that concentrations will be within limits in the receiving body of water before a discharge permit is issued. This documentation can be in the form of experimental results or mathematical predictions. In either case, accepted procedures must be followed and the models used must be approved.

This book concentrates on mathematical modeling of environmental discharges. It is intended to be a reference book for practicing engineers and scientists involved in the permitting process and as an advanced level text book for seniors or first-year graduate students in engineering or environmental science.

The first chapter discusses the fundamentals of turbulent jet mixing, dilution concepts, and mixing zone concepts. The second chapter is devoted to diffuser configurations and head loss calculations. The remaining chapters discuss different modeling techniques and a few presently accepted models that use these concepts. The models are discussed in detail giving theoretical background, restrictions, input, output, and examples of use. Chapter 3 discusses Lagrangian integral methods and the EPA UM two-dimensional diffuser model. It also discusses the PLUMES interface used to create the universal data file used by several models for input and an introduction to the Windows-based version, WISP. Chapter 4 discusses Eulerian integral methods, the EPA UDKHG (UDKHDEN), three-dimensional diffuser model, and PDSG (PDS) surface discharge model. Chapter 5 discusses empirical techniques, the RSB diffuser model, and the CORMIX family of models for both diffusers and surface discharge. Chapter 6 is a brief discussion of atmospheric models including Gaussian puff and integral cooling tower plume models. Chapter 7 is a discussion of accuracy, numerical models, and recommendations. Numerous case studies and examples are considered using each model. Problems have been included in each chapter that are characteristic of those encountered by the author.

Most of the models discussed are available in the Internet. Users will be able to download each for their use. The case studies included in the book are complete and do not require the user to have the models.

The author is appreciative of Dr. Walter Frick at the U.S. EPA Athens, Georgia CEAM center, Dr. Robert Doneker at the Oregon Graduate Institute, Dr. Phil Roberts at Georgia Tech. University, and Dr. Gehardt Jirka previously of Cornell University for the many discussions we have had over the years regarding the development of the CORMIX, RSB, and PLUMES models.

Table of Contents

1 Fundamentals

1.1 PHYSICAL CONCEPTS IN TURBULENT JET MIXING

1.1.1 ENTRAINMENT AND DIFFUSION

Entrainment is the process of drawing ambient fluid into a jet that is discharged into the ambient. It can be described by considering the interaction between two adjacent fluid layers, one going faster than the other. Several mechanisms cause the faster fluid to accelerate the slower fluid. One is viscous shear between the two layers where the slower fluid is simply pulled along by viscosity by the faster fluid making it become part of the jet. A larger effect is usually due to turbulence. Figure 1.1 shows a typical turbulent jet illuminated by fluorescent dye. The turbulent eddies at the edge of the jet grab large quantities of ambient fluid and carry it along with the jet. This turbulent action also stirs the two fluids together, with the ambient fluid penetrating the discharged fluid and the discharge fluid penetrating the ambient.

If the discharged fluid is less dense than the ambient, a buoyant jet results. In addition to viscous and turbulent shear, buoyant forces cause the plume to rise relative to the ambient. This induces a secondary motion of the discharged fluid relative to the ambient. This secondary motion also entrains ambient fluid into the jet by viscous and turbulent shear.

The net result of jet and buoyant entrainment is a jet whose mass flow rate increases as it entrains more and more ambient fluid. Since mass must be conserved, ambient fluid that is entrained into the jet must be replaced by other fluid from the ambient. This causes a net flow of fluid from the ambient toward the jet. This entrainment process is a result of a momentum exchange between the jet and ambient. Near the discharge, the entrainment rate is high where the relative velocity between the jet and ambient are high. The entrainment rate decreases, however, as the jet penetrates the ambient and it loses momentum to the ambient fluid it has entrained.

Eventually the momentum of the jet is lost to the ambient. When this happens, further mixing of the ambient and the discharge fluid occurs by ambient turbulent mixing and diffusion. Ambient turbulence causes mixing at the edge of the plume the same as jet turbulence did earlier but usually at a much slower rate, since there is no relative motion between the discharged fluid and ambient. The discharged fluid flows freely with the ambient. This crossover from jet-induced mixing to ambient turbulent mixing occurs gradually as the plume develops from a jet. The relative importance of the two determines the boundary between the near field and the far field. The **near field** is where jet mixing dominates and the **far field** is where ambient turbulent mixing dominates.

Diffusion of the discharged fluid into the ambient is caused by molecular diffusion and turbulent mixing. Molecular diffusion results from the random motion

FIGURE 1.1　Illuminated turbulent jet.

of molecules in a fluid. In a fluid with a uniform concentration of a particular tracer or pollutant, random motion causes the same number of molecules to cross an imaginary boundary in both directions. If the concentration of the tracer is higher on one side of the boundary than on the other, more molecules of this tracer will cross to the low concentration side than in the reverse direction. This results in a diffusion of tracer particles from the high concentration region into the low concentration region. The rate of diffusion is generally expressed by Fick's Law:

$$\bar{f}_i = -D_i \nabla C_i \qquad (1.1)$$

where $\bar{f}$ is the vector diffusion flux, D_i is the diffusion coefficient of tracer i, and ∇C_i is the gradient of the concentration of tracer i. The minus sign is included to give a positive diffusion flux in the direction of a decreasing concentration. The diffusion coefficient is often interpreted as the combination of both molecular and turbulent diffusion. Molecular diffusion is usually negligible relative to turbulent diffusion.

1.1.2　Development Zone

The turbulent velocity and concentration profile in a duct are blunt. As a result, profiles at the exit of a discharge pipe are nearly "top hat" as shown on Figure 1.2. Once the jet comes in contact with the ambient, the diffusion process produces a shear layer where the ambient penetrates the jet causing it to slow down and the jet penetrates the ambient causing it to speed up. Eventually the ambient penetrates the

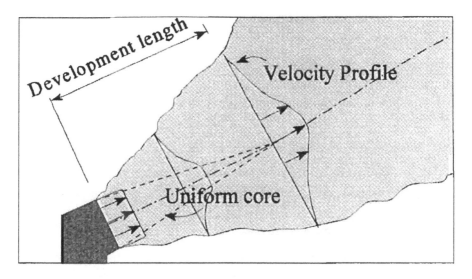

FIGURE 1.2   Development zone profiles.

jet to the center. The centerline velocity and concentrations do not change until the shear layers meet at the center. This region where the profiles change from "top hat" to bell shaped is known as the zone of flow establishment. Beyond the zone of flow establishment is the zone of established flow where the profiles are all similar changing only in width and centerline magnitude. In this region profiles have been found to be expressed by the Gaussian probability function.[1]

$$\frac{u}{u_{cl}} = e^{-(r/b)^2} \tag{1.2}$$

where u_{cl} is the centerline velocity. A close approximation to this function is the 3/2 power low profiles given by

$$\frac{u}{u_{cl}} = \left[1 - \left(r/b'\right)^{3/2}\right]^2 \tag{1.3}$$

Notice that plume radius b' in Equation (1.3) is the full radius while b in Equation (1.2) is a partial plume radius given by the standard deviation of the profile. Figure 1.3 shows how these two profiles compare when plotted against r/b' with $b = 0.534\ b'$. This forces the two profiles to cross at $u/u_{cl} = 0.5$.

1.1.3   Jets and Plumes

Often the terms *jets* and *plumes* are used synonymously. They do, however, have slightly different meanings. The term "jet" implies a high velocity relative to the ambient, whereas a "plume" has little or no momentum but spreads like a billowy cloud or rises due to buoyancy only. In reality, most discharges from a point source

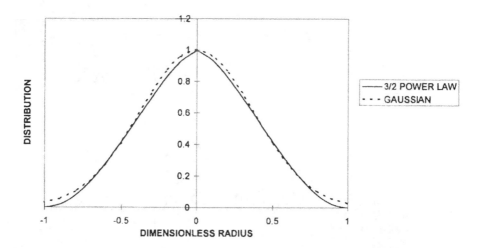

FIGURE 1.3 Gaussian and 3/2 power law profiles.

start as a jet but rapidly loose their momentum to the entrained ambient and become a plume a short distance from the discharge.

Often the ratio of the momentum of a jet to its buoyance is used as a scaling parameter. This ratio in dimensionless terms is the densimetric Froude number. Considering discharge conditions this can be expressed as

$$F_o = \frac{U_o}{\sqrt{\dfrac{\Delta\rho}{\rho}gD_o}} \tag{1.4}$$

where $\Delta\rho$ is the density deference between the ambient and discharged fluid, U_o and D_o are the discharge velocity and diameter, respectively. Discharges with Froude numbers much greater than one are considered momentum jets, while those with Froude numbers only slightly greater then one are termed buoyant jets. Discharges with Froude numbers less then unity have no jet phase at all and result in buoyant plumes.

This text is mainly concerned with the region close to the discharge, or near field where the discharge changes from a jet to a plume. The term "jet" will refer to that region very close to the discharge. The term "plume" will refer to the rest of the field of interest.

1.2 DILUTION

1.2.1 DILUTION AND CONCENTRATION

Dilution is often used to determine how much ambient fluid has been entrained into the plume. The usual definition of **dilution** is the ratio of flow in the plume to the discharge flow, Q/Q_o. Since the velocity varies over the area of a plume, the dilution for a single round jet is calculated from

$$S = \frac{\int\limits_{0}^{\infty} u dA}{U_o \pi \dfrac{D_o^2}{4}} \qquad (1.5)$$

where the symbol S is used for dilution. With this definition, a dilution of 10 comes from a plume that consists of 9 parts entrained fluid and 1 part discharged fluid.

The conservation of mass for a particular tracer can be given by

$$Q_o C_o + E C_a = \int\limits_{0}^{\infty} u C dA \qquad (1.6)$$

where C_o is the mass concentration of the tracer in the discharge, C_a is the concentration of the same tracer in the ambient fluid, and E is the total entrained fluid to the point in the plume where the integral is evaluated. This equation assumes that the densities are all nearly the same and cancel. If the ambient concentration is zero, Equation (1.6) is often simplified to $Q_o C_o = Q \overline{C}$ where $\overline{C}$ is the flux average concentration across the plume. This expression leads to

$$\frac{\overline{C}}{C_o} = \frac{1}{S} \qquad (1.7)$$

or the ratio of the average concentration in the plume to the discharge concentration is equal to one over the dilution.

If the ambient concentration is not zero, the above simplification cannot be used. Instead, Equation (1.6) is simplified to

$$Q_o C_o + \left(Q - Q_o\right) C_a = Q \overline{C} \qquad (1.8)$$

which can be rearranged to give

$$\frac{\overline{C}}{C_o} = \frac{1}{S} + \frac{(S-1)}{S} \frac{C_a}{C_o} \qquad (1.9)$$

A careful inspection of the derivation of these equations will show that the definition of the flux average concentration $\overline{C}$ when using the 3/2 power law profiles on a single circular jet for both velocity and concentration can be expressed as

$$\overline{C} = C_{cl} \frac{\int_0^{b'} \left[1 - \left(r/b'\right)^{3/2}\right]^4 r dr}{\int_0^{b'} \left[1 - \left(r/b'\right)^{3/2}\right]^2 r dr} = 0.519 C_{cl} \qquad (1.10)$$

where C_{cl} is the centerline concentration of the species of concern. Using this result leads to the expression

$$\frac{S}{S_{cl}} = 1.93 \qquad\qquad (1.11)$$

which is the ratio of the flux average dilution to the centerline or minimum dilution for a single circular plume. When Gaussian profiles are assumed, the ratio of average dilution to centerline dilution is 2.0. For completely merged line plumes, the ratio is 1.43 for the 3/2 power law profile.

1.3 REGULATIONS

The concentrations of toxic and thermal pollutants within the environment are carefully controlled by regulatory agencies. Whenever there is a municipal or industrial discharge of fluids into the environment, those responsible must demonstrate that the discharge will not increase concentrations of toxic materials or temperature above specified levels. As a result of the Clean Water Act,[2] concentrations of some pollutants are allowed above the critical values within a specified "mixing zone." In order to minimize the impact of these pollutants on the environment, the allowable concentrations at the edge of the mixing zone and the size of the mixing zone itself are constrained to small values. The actual values vary from site to site and are set by state and federal regulatory agencies.

Pollutants are categorized according to (1) conventional or naturally occurring, (2) unconventional, (3) toxic, (4) heat, and (5) dredge spoils. An example of a conventional pollutant would be biological oxygen demand (BOD). Unconventional pollutants are those that can't be categorized as naturally occurring or toxic; chemical oxygen, for example. A special class of pollutants similar to dredge spoils is drilling mud.

The U.S. Environmental Protection Agency (EPA) has set two water quality criteria for toxic pollutants. One is the criterion for continuous concentration (CCC) or chronic concentration and must be met within a regulatory mixing zone. The second is the criterion for maximum allowable concentration (CMC) or acute concentration. This maximum allowable concentration is very restrictive and must be met within a small toxic dilution zone. In some states this toxic dilution zone is often called the zone of initial dilution or ZID. As a result, the zone of initial dilution and mixing zone are usually different, with the zone of initial dilution being much more restrictive and smaller.

This toxic mixing zone or zone of initial dilution has been defined by the EPA as the smaller of the following:

1. 10% of the distance to the regulatory mixing zone.
2. A distance in any direction equal to 50 times the square root of the discharge port area.
3. A distance equal to 5 times the local water depth.

Each state can impose the EPA's regulations or specify more stringent regulations of its own. Since the regulatory mixing zone can be set to whatever the state feels is appropriate, condition (1) above is arbitrary. Conditions (2) and (3) are more physical and insure sufficient distance for mixing without interaction with the surface or bottom. Additional conditions may also be imposed, such as not allowing the plume to encompass more than 25% of a given stream or river within the mixing zone or that the discharge velocity must be at least some minimum value. Earlier EPA regulations required a discharge velocity of 3 m/s (10 ft/s), but this was found to be unreasonable with discharges that ranged from summer low conditions to high storm conditions. This restriction has been relaxed in more recent regulations.[3] One way of keeping the discharge velocity up at lower discharge rates is to use flexible rubber "duck bill" nozzles on the discharge ports. These rubber check valves also prevent intrusion of sand, salt water, and biolife at low discharge rates.

Allowable concentrations of pollutants are often determined by bioassay. In this analysis, specified species such as fish minnows are exposed to various concentrations of the pollutant for a period of time (24 hours, 96 hours, etc.). The concentration that causes a certain percentage of the species being tested to die after the given exposure time is called the lethal concentration. For example, the concentration that kills 50% of the test species after 96 hours is called the 96 hour LC50.

Each state sets its own limits, but the 96 hour LC50 is often used to determine the **acute** toxicity for the toxic mixing zone or CMC. This value decreases as toxicity increases and has led to some confusion. As a result, Toxic Units are defined as 100 divided by the lethal concentration. This unit increases as toxicity increases. Thus, Acute Toxic Unit or TUA is often defined as 100/LC50.

Chronic concentrations are determined from the maximum concentration in a bioassay that shows *no* noticeable effect on the test species after the specified period of time or LC0. This concentration gives the CCC for the mixing zone. Again, toxic units for the chronic conditions are defined as 100/LC0.

For municipal ocean discharges, the EPA has defined a mixing zone or zone of initial dilution (ZID)[4] as that lateral distance to where the plume either reaches the surface or is trapped by stratification. Unfortunately, this simple definition cannot easily be used with river or lake discharges where the more complicated definitions arise. The terms "zone of initial dilution" and "mixing zone" are also used for river and lake discharges, but their meanings are for the acute and chronic concentrations, respectively. For example, a mixing zone for a river may be defined as a 100 ft semicircular region downstream of an outfall with its origin at the outfall. Where tidal effects are present, the mixing zone may be defined as a rectangular region whose length equals the diffuser length plus 100 ft with the width extending 300 ft downstream and 50 ft upstream. The ZID or toxic mixing zone may be 10% of these values or shorter as defined by water depth or discharge port area criteria. These definitions are site-specific but usually follow specific guidelines set up by the state. Some people define an **initial mixing zone** as that region where small-scale turbulence induced by the jet causes rapid mixing. The region ends when this turbulence is suppressed by plume collapse. This is the mixing zone used in the RSB model discussed in Chapter 5.

In some cases, no numerical mixing zone may apply. In these cases, an applicant may be required to specify where the critical concentrations are expected to occur. The state may accept these distances as the effective mixing zone or it may ask the applicant to come up with alternate discharge configurations that will decrease these distances. This may continue until the applicant comes up with a design that gives expected distances to achieve critical conditions that the state can accept.

1.4   AMBIENT CONSIDERATIONS

1.4.1   DEPTH

Ambient conditions play an important role in the concentrations of pollutants that occur at the edge of the mixing zone. These include receiving water depth, current, density stratification, tidal fluctuations, channel configuration, and bottom geometry. In general, the deeper the discharge, the better chance there is of meeting regulations. This is because more water is available for dilution, there is less chance of impingement with the bottom and surface, and the trajectory of the plume is longer before reaching the surface. This often requires that the outfall be placed further out into the ambient receiving water where the water is deeper.

Discharge into shallow water is always a problem. The plume quickly fills the water column. When the plume hits the bottom or surface of the receiving water, the rate of dilution decreases, with entrainment only coming from the sides of the plume in contact with fresh water. In addition, the plume may scour the bottom. Modeling this type of plume is very difficult. Methods of approximating special cases are discussed in Chapters 3 and 4.

Figures 1.4, 1.5, and 1.6 illustrate several types of surface and bottom interaction. Figure 1.4 shows a deep water buoyant where the plume rises to the surface. It then spreads out on the surface as a surface plume with little lateral momentum being carried away by ambient surface currents. There is no bottom interaction with this type of discharge. Figure 1.5 shows a near vertical discharge into shallow water with little or no ambient current. The strong interaction with the surface causes an unstable near field with the plume entraining itself. There is very little dilution. Figure 1.6 shows a near horizontal discharge into a shallow flowing ambient such as is encountered in many river and stream diffusers. The current and discharge momentum carries the plume downstream. The vertical discharge momentum and plume buoyancy cause the plume to rise. Because of the shallowness of the river, the plume quickly fills the water column with the entrainment impeded where these interactions occur. Due to the Coanda effect, the plume actually sucks itself to the bottom, causing the bottom interaction to occur before it might be expected.

1.4.2   STRATIFICATION

Ambient density stratification can have a profound effect on plume dynamics. Figure 1.7 shows a typical plume from the discharge of a buoyant jet into a stably stratified ambient. As the jet mixes with the ambient, the plume's density increases, gradually approaching that of the ambient. The plume momentum and positive

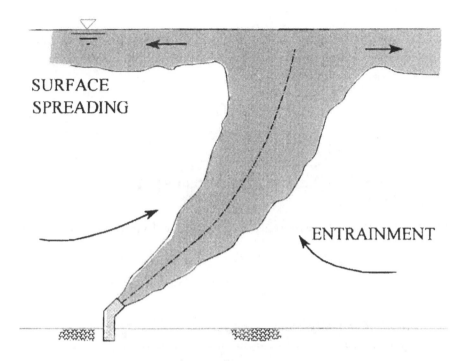

FIGURE 1.4   Deep water discharge with surface spreading.

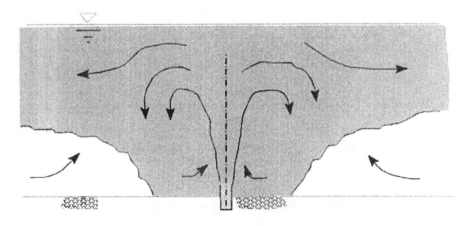

FIGURE 1.5   Near vertical shallow water discharge with unstable flow field.

buoyancy cause the plume to rise. As it rises, it encounters an ambient whose density is decreasing. If the plume continues to rise, the density of the plume will reach that of the ambient. If the plume continues to rise due to its momentum, it encounters an ambient whose density is less than its own. This causes negative buoyancy, which stops the plume's rise and causes it to drop back down toward natural buoyancy. It

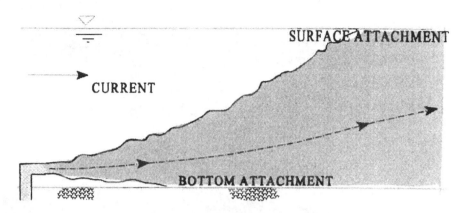

FIGURE 1.6   Near horizontal discharge in shallow water with surface and bottom interaction.

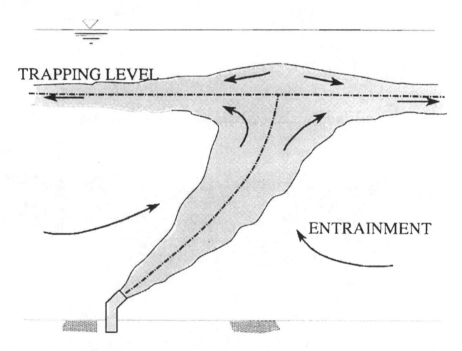

FIGURE 1.7   Plume trapping due to density stratification.

may actually overshoot in the opposite direction, oscillating about the neutral buoyant level until the vertical momentum has dissipated. The eventual neutral buoyancy
level is call the "trapping level" and the plumes is "trapped."

Even the slightest ambient stratification can trap a plume, but conditions where
the ambient density is layered with abrupt changes in density always trap the plume.
This layer where the ambient density suddenly changes is called the **pycnocline**.
This phenomenon often occurs in lakes due to thermal stratification, and in estuaries

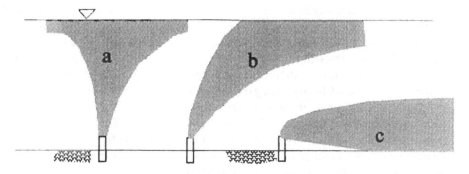

FIGURE 1.8 Plume possibilities with various current velocities for vertical discharge:
(a) low current, (b) moderate current, (c) high current.

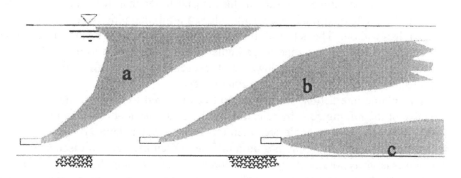

FIGURE 1.9 Plume possibilities with various current velocities for horizontal discharge:
(a) low current, (b) moderate current, (c) high current.

or coastal regions due to salinity stratification. Once the plume is trapped, the dilution
process is greatly reduced with the layered plume spreading out in a layer below
the surface, with mixing occurring primarily due to ambient turbulence and diffusion.

The actual density profile in the ambient is important in determining mixing
zone concentrations. Linear approximations to a variable density profile often lead
to false trapping levels that cause predicted dilutions to be in error. Earlier dilution
models such as DKHPLM[5] always assumed linear stratification and were found to
be in error in many stratified cases. Later versions of this model discussed in
Chapter 4 have corrected the problem.

1.4.3 AMBIENT CURRENT AND DISCHARGE DIRECTION

The ambient current affects the plume in a variety of ways. It eventually always carries
the plume in the direction of the current regardless of the direction of the discharge.
Figures 1.8 and 1.9 show several plume shapes that occur depending on the magnitude
of the current. Figure 1.8 is for vertical discharge and Figure 1.9 is for horizontal
discharge. For no ambient current, a buoyant plume will rise all the way to the surface

or trapping level and spread out in all directions. Entrainment is due to jet induced entrainment and plume buoyancy only for either vertical or horizontal discharge.

At low to moderate currents and vertical discharge, the ambient current forces fluid into the plume through any surface of the plume that it can see. If the momentum of the jet is sufficient, the current will cause two vortices to form on the back side of the jet as is found behind pilings and columns. These vortices also cause the ambient to be mixed with the plume. The momentum of the entrained fluid causes the plume to rapidly bend over. This near field interaction between the ambient current and plume causes very rapid mixing. Once the plume bends over, however, the entrainment is a function of the relative velocity between the plume and ambient current and the dilution rate can drop off quickly.

For horizontal discharges, the current can actually decrease the dilution rate by decreasing the relative velocity between the discharge and current. In fact, if the current and discharge velocities are equal, mixing is by turbulence only. Since the current carries the plume downstream further and faster than it would without the current, the discharged fluid reaches the edge of a mixing zone in less time with a current than without. The net result is that the ambient current can cause the dilution at the edge of the mixing zone to be higher in some cases due to the high plume interaction in the near field, and lower in others because of reduced relative velocity and reaching the edge of the mixing zone faster.

This plume interaction with the current can be used to increase the dilution at the edge of the mixing zone by selecting discharge angles that cause an interaction. Limits on the discharge angle are often dictated by water depth, river width, and interaction of neighboring plumes on a diffuser. Often the vertical discharge angle is made as large as possible from the horizontal without causing the plume to surface too soon. In shallower water, the discharge angle is made just high enough to keep the plume off the bottom. Horizontal discharge angles can vary from downstream to cause this interaction. For single port discharges, there is no limit on the angle, but for line diffusers, shielding by upstream ports reduces the interaction, as shown on Figure 1.10. The upstream discharge blocks the current from interacting with downstream ports until the upstream plumes bend over into the path of the downstream plumes. This effect can clearly be seen when the wind blows along a line connecting the plumes from multiple port mechanical draft cooling towers.

Current also affects the dynamics of the plume. The two vortices discussed above that occur when there is an interaction between the current and jet cause the plume to divide into two regions. Figure 1.11 shows the cross section of the plume generated by a vertical jet discharged normal to the ambient current. Plume illumination was produced using laser fluorescent dye and a light slit. These vortices cause the concentration and velocity profiles within the plume to be bimodal. In some cases, such as in a natural draft cooling tower or smoke stack with moderate to high wind, the plume will actually bifurcate and become two separate spinning plumes. As a result, the Gaussian and 3/2 power law profiles used in many models are not very accurate in describing details within the plume under these conditions.

Many regulatory agencies limit the current that can be used in a mixing zone analysis. Since it is generally considered that current will increase near field mixing for most discharge conditions, worst-case conditions are when the current is low.

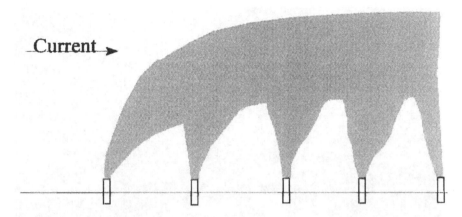

FIGURE 1.10 Discharge from a line diffuser parallel with the current with shielding from upstream ports.

As a result, some agencies will only accept zero current predictions; however, most agencies allow more realistic values. Typical low current conditions used are either the lowest 10% current expected or 7Q10 river conditions. **7Q10 river conditions** are defined as the lowest 7-day average river flow that occurs over a 10-year period.[3]

1.4.4 TIDES

Tides cause both the depth, current, and current direction to vary with time. In addition, since most tidal concerns are in coastal areas, tides also change the stratification as salt water moves in and out of bays and rivers. Since most of the currently used mixing zone models are steady state, they are unable to predict tidal conditions. Luckily, most mixing zones are small enough that discharged fluids reach the edge within minutes of being discharged. During this time, tidal changes are minor. As a result, several steady-state runs can be made with different ambient conditions during a tidal cycle and yield reasonable predictions. The main problem with this approach is that in some configurations, the incoming tide causes a flow reversal that causes the background concentrations of discharge pollutants to vary with time. The COR-MIX model discussed in Chapter 5 approximates this process for a short period.

If variable background concentrations must be accurately accounted for, transient numerical models must be used. They vary in complexity from simple one-dimensional models to full finite-element models. The WASP5 EPA model can be downloaded from the EPA bulletin board in Athens, Georgia.*

1.4.5 CHANNEL CONFIGURATIONS

Channel configurations are only handled indirectly in most presently used mixing zone models. In several of the models to be discussed later, such as PLUMES-UM, UDKHDEN, and PDS, the user must manually determine if the plume interacts with

---

* WASP can be downloaded from the Internet at ftp://ftp.epa.gov/epa_ceam/DOS

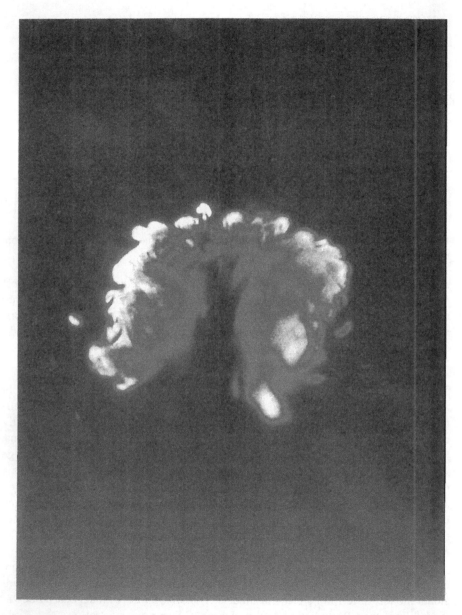

FIGURE 1.11 Illuminated cross section of a plume discharged at right angles to the ambient current showing horseshoe shape and plume vortices.

channel banks. PLUMES indicates when bottom attachment occurs but does not give accurate predictions beyond this point. UDKHG graphically shows bottom effects. The CORMIX models keep track of both bank and bottom interaction, but the channel is assumed to be rectangular and not varying downstream. Since most river mixing zones are usually only a few hundred feet long, or less, channel variations can often be ignored, and manually keeping track of bank interaction is sufficient.

2 Diffusers

2.1 CONCEPTS

The main objective of a diffuser is to spread the effluent within the ambient in order to achieve more rapid mixing. A large number of small diameter ports has a larger contact area with the ambient than just a single port with the same discharge velocity and flow. For example, a 1-m diameter discharge has a 3.14-m circumference while 100 10-cm diameter ports having the same area have a total circumference of 31.4 m. As a result, the initial mixing rate would be approximately 10 times faster in the 100-port diffuser than from the single port.

This assumes, of course, that ambient fluid is free to flow to each jet without interference from other jets. In the actual case, considerable interference can take place depending on spacing, orientation of jets of the diffuser, and discharge direction relative to the current. Figure 2.1 shows a typical in-line diffuser with the plume divided into zones of consideration. The zone of flow establishment is where "top hat" profiles change to Gaussian profiles. In this region, centerline concentrations, temperatures, and velocities do not change. In the fully developed single plume zone, conditions are essentially the same as for each individual plume as if it were alone except as discussed below. In the merging region, plume edges grow into each other, causing concentrations, temperatures, and velocities at the interface to increase toward plume centerline values. Since plumes are merging along the diffuser centerline, ambient fluid is prevented from mixing with each plume along this centerline. This reduces the rate of entrainment. Eventually when the plumes are completely merged, they resemble a line plume.

Mutual attraction causes the plumes to bunch together. This can be explained by considering the two outermost plumes. The second one in entrains fluid from all sides, setting up currents toward it. Since the last plume on the edge is in that flow field, it is gradually pulled toward the second plume in. These two, in turn, are pulled toward the third one in and so forth. This causes the plumes to merge faster than would be normally predicted. Figure 2.2 shows this effect on a diffuser with ports on both sides. Since the plumes from the two sides attract each other and eventually merge together, alternating diffuser ports is not as effective as might be expected. A conservative assumption is that they are all on the downstream side of the diffuser.

Figure 1.10 in the previous chapter shows a diffuser oriented parallel to the current with ports discharging out into the current. The upstream plume sees the full effect of the current, but it shields the current from the downstream ports until it has bent over into their path. As a result, the current's interaction with all ports is limited in the near field with this orientation. Entrainment of the downstream ports is mainly by jet-induced entrainment. And again, once the plumes have merged

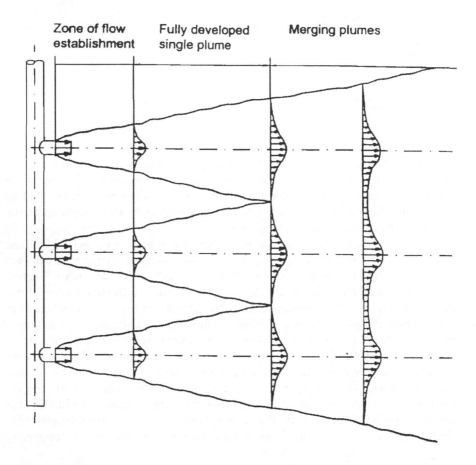

FIGURE 2.1   Typical in-line diffuser with the plume divided into zones.

together, they become one big plume, with the effects of the multiple port discharge disappearing.

Diffuser orientations between parallel to the current and perpendicular to the current will have some current interaction with each port, to a lesser extent than for the perpendicular diffuser but more than if the interaction were ignored altogether. This issue will be discussed more in later chapters.

One approach to approximate line diffusers is to use an **equivalent slot concept**. In this approximation, the discharge from a line of equally spaced ports is assumed to be the same as from a line slot discharge that has the same length, discharge flow, and momentum as the multiple ports. This concept is useful in analyzing the characteristics of certain multiport discharges, but as with most simplified analyses, the limits of use should be understood. The fundamental premise behind the equivalent slot is that beyond the point of merging, the plumes from a row of equally spaced jets become the same plume as the one generated by an equivalent slot. This presumes that up to the point of merging, the mixing mechanisms of the two are the same, and of course they are not. The effect of this difference on the ultimate dilution

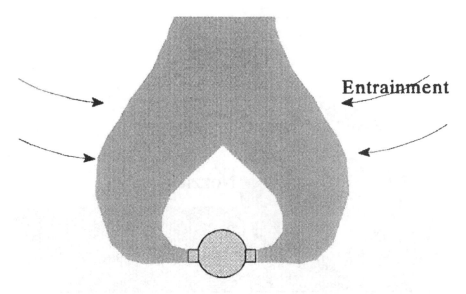

FIGURE 2.2   Resulting plume from a diffuser with ports on both sides.

would be minor if the major portion of the dilution occurred after merging. This would either be for large distances away from the point of discharge or for closely spaced discharge ports. Liseth[6] suggests that the equivalent slot concept should only be used at distances of 80 times the spacing between ports and beyond. For ports that are spaced 10 discharge diameters apart, this distance is 800 discharge diameters downstream where very large dilutions usually exist. If interest is in the near field, the equivalent slot concept should probably not be used. Unfortunately, many models such as CORMIX2 and RSB use the equivalent slot extensively.[7]

2.2   GEOMETRY

Diffusers consist of a manifold pipe, closed at the end, with discharge ports spaced along its length. The ports can be simply holes in the side of the manifold as shown on Figure 2.3 or fitted with risers as shown on Figure 2.4. The risers may have one or more discharge ports. Risers are used to raise the discharge point above the manifold, to allow the manifold to be buried or to allow check valves to be used to prevent back flow. One of the check valves that is becoming more popular with diffuser nozzles is the rubber "duck bill" type check valve made by the Red Valve Company,* since they do not stick or foul like mechanical swing check valves do.

The three major diffuser configurations are unidirectional, staged, and alternating, as shown in Figure 2.5; however, many variations of these exist. Unidirectional diffusers are common in rivers where residual currents do not reverse. The fanned design tends to counter the mutual attraction of jets as discussed above and increases near field mixing over unfanned diffusers. The "V" design unidirectional diffuser is

---

* Red Valve® Company, Inc., P.O. Box 548, Carnegie, PA, 15106.

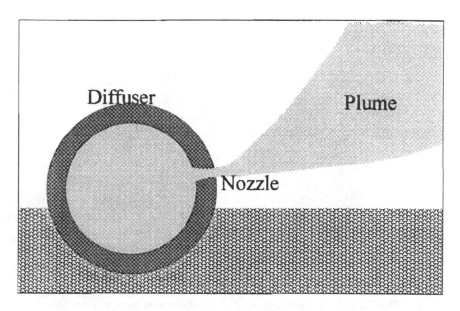

FIGURE 2.3   Diffuser manifold with discharge nozzles.

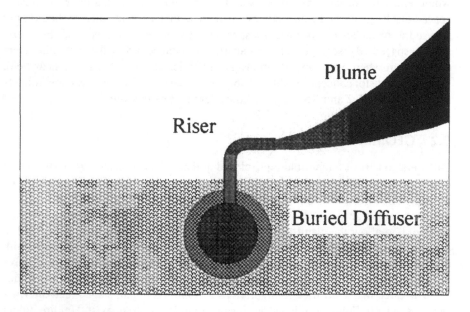

FIGURE 2.4   Diffuser with risers to have discharge above manifold.

often used when it is cheaper to have fewer risers, spaced further apart but with two or more ports on each riser. Staged diffusers usually have ports on both sides that point generally along the diffuser. Values of β up to about 20° are used. Because of

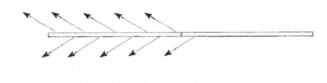

Staged with
alternating ports

Unidirectional

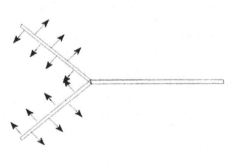

Y diffuser with
alternating ports

FIGURE 2.5 Unidirectional, staged, and alternating arrangements.

their port orientation, staged diffusers set up currents and circulation patterns along the diffuser line forcing ambient fluid in the direction of the ports. Although the plumes quickly interfere with each other, they do set up large lateral entrainment velocities toward the diffuser.

Alternating port diffusers are often used in bays and coastal regions where currents vary in direction and where diffuser length is of major concern. Since the net lateral jet momentum is zero, buoyancy and currents play important rolls. Accurate modeling of staged and alternating diffusers is difficult due to mutual attraction and merging. Conservative estimates can be obtained by assuming all ports are on the downstream side of the diffuser while optimistic predictions can be obtained by treating each side independent of the other. Actual dilutions will be somewhere between the two limits.

Parameters to be considered when designing a diffuser are as follows:

1. A large number of small ports gives better dilution than the opposite.
2. The minimum port diameter is usually controlled by jet velocity requirements, head loss considerations, and problems with clogging.
3. The wider the spacing between ports, the closer the dilution rate will approach a single port discharge for each nozzle. Wide spacing, however, leads to long, costly diffusers. Ports are usually placed as close together as is possible while still satisfying regulations.

2.3 DIFFUSER HEAD LOSS CONSIDERATIONS

2.3.1 FUNDAMENTALS

One of the important considerations in diffuser design is the overall head loss within the diffuser and the flow distribution between ports. It is desirable to have the head loss within the diffuser as low as possible so that available upstream heads are sufficient during peak flow conditions to prevent flooding in gravity flow systems or to keep pumping power down in pumped systems. At the same time, it is desirable to have high port discharge velocities for rapid mixing. High diffuser manifold velocities keep solid material suspended but cause high pressure variations along the manifold, which results in poor flow distribution along the length of the diffuser. Large diameter diffuser manifold pipes with low fluid velocities give good flow distribution between ports but are more costly than smaller diameter pipes. The actual design of the diffuser is usually an iterative process with compromises made between the different concerns.

The equations relating the head loss along a diffuser manifold couple the energy equation along the pipe and flow out the diffuser ports.[8] Considering Figure 2.6, which shows two adjacent ports in a diffuser, the energy equation for flow between the ports can be expressed as

$$Z_n + \frac{p_n}{\gamma} = Z_{n-1} + \frac{p'_{n-1}}{\gamma} + h_n \tag{2.1}$$

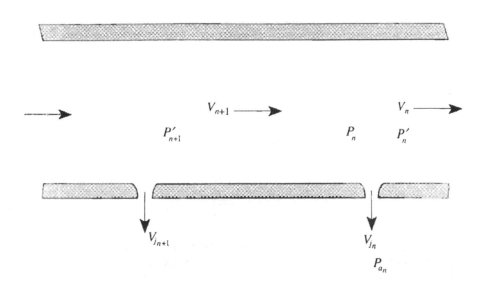

FIGURE 2.6 Defining sketch of two adjacent ports for energy equation considerations.

where h_n is the head loss in the manifold between the ports and can be calculated using any of several methods, including Manning's equation, the Hazen Williams equation, the Swamee–Jain equation, or from

$$h_n = f_n \frac{L}{D} \frac{V_n^2}{2g} \tag{2.2}$$

where f_n is the Darcy–Weisbach friction factor and can be found from the Moody diagram giving friction factor as a function of pipe Reynolds number and the relative roughness of the pipe or from the Colebrook–White equation.[9] The Swamee–Jain equation[10] has been found to closely approximate the friction factors found in the Moody diagram and gives head loss directly. It can be expressed as

$$h_n = \frac{0.203 \dfrac{Q_n^2}{gD^5} L}{\left[\log_{10}\left(\dfrac{\varepsilon}{3.7D} + \dfrac{5.74}{R^{0.9}} \right) \right]^2} \tag{2.3}$$

where ε is the absolute roughness of the pipe, Q_n is the pipe flow, R is the Reynolds number of the water in the pipe $V_n D/\nu$. The kinematic viscosity of water, ν, is a function of temperature but can be taken as 1.31×10^{-6} m²/s (1.41×10^{-5} ft²/s) for most outfalls. This equation has been found to be within about 1.0% of the values given by the Colebrook–White equation.

The energy equation relating flow from the manifold pipe out diffuser port n can be expressed as:

$$\frac{P_n}{\gamma} + \frac{V_{n+1}^2}{2g} = \frac{P_{a_n}}{\gamma} + \frac{V_{j_n}^2}{2g} + k\frac{V_{j_n}^2}{2g} \tag{2.4}$$

where k is the loss coefficient for the discharge port. The parameter p_{a_n} is the ambient pressure into which port n discharges and is a function of the port depth and ambient density. Solving (2.4) for the jet velocity yields

$$V_{j_n} = \sqrt{\frac{2gE_{j_n}}{1+k}} \tag{2.5}$$

where E_{j_n} is the net available head across the port and represents the difference between the pressure head on the outside caused by the port depth and the total head inside the pipe at this port.

$$E_{j_n} = \frac{P_n}{\gamma} + \frac{V_{n+1}^2}{2g} - \frac{P_{a_n}}{\gamma_a} \tag{2.6}$$

The flow rate through the ports is usually expressed in terms of the available head

$$Q_n = C_n A_n \sqrt{2g\,E_{j_n}} \tag{2.7}$$

where C_n is the port discharge coefficient and is a combination of the minor loss coefficient and the nozzle contraction coefficient. Empirical expressions for C_n can be expressed as*

$$C_n = 0.975\left(1 - \frac{V_{n+1}^2}{2gE_{j_n}}\right)^{3/8} \tag{2.8}$$

for bell-mouth ports and

$$C_n = 0.63 - 0.58\frac{V_{n+1}^2}{2gE_{j_n}} \tag{2.9}$$

for sharp-edged ports.

If the diffuser is equipped with risers, as shown on Figure 2.4, the head loss can be found by adding up all the minor losses within the riser and adding the contraction

---

* Raun, A.M., Bowerman, F.R., and Brooks, N.H., Diffusers for dispersal of sewage in sea water, *Journal of the Sanitary Engineering Division*, ASCE, Vol. 86, No. SA 2, 1960.

coefficient. It is usually assumed that the total head is constant on either side of port n such that

$$\frac{P_n}{\gamma} + \frac{V_{n+1}^2}{2g} = \frac{P_n'}{\gamma} + \frac{V_n^2}{2g} \tag{2.10}$$

Since the right side of Equation (2.10) is known and p_a can be found from the port depth and the ambient fluid density profile, the left side of Equation (2.10) along with p_a/γ gives E_{j_n} from Equation (2.6). The value of the discharge coefficient, however, requires the knowledge of V_{n+1}, which is not known until Q_{j_n} is found using the discharge coefficient. This dilemma is solved by using an iterative approach. As a first guess, the discharge coefficient is evaluated using E_{j_n} and V_n. The port flow is calculated from Equation (2.7). The velocity V_{n+1} is calculated from the flow downstream of the port added to the port flow divided by the area in the manifold upstream of the port,

$$V_{n+1} = \left(\frac{D_n}{D_{n+1}}\right)^2 V_n + \frac{Q_n}{\left(\frac{\pi}{4} D_{n+1}^2\right)} \tag{2.11}$$

This value of V_{n+1} is used to update C_n, and the process is repeated until the value of V_{n+1} does not change. Since calculations must start with known values, the usual procedure is to assume an E_{j1} for the last port of the diffuser. Calculations proceed to the upstream end of the diffuser where the total flow calculated is compared to the desired flow. If the two flows do not match, a new E_{j1} is selected and the process is repeated until the calculated flow and desired flow are equal.

2.3.2   THE PLUMEHYD PROGRAM

Performing the calculations to determine the total head loss in a diffuser is best accomplished on a computer that can do the iterations rapidly. Several programs exist in the open literature that perform these calculations. The one most readily accessible is the PLUMEHYD program that comes with the EPA's PLUMES package.* This program is good for diffusers without risers or where the losses in the risers are negligible.

Input to PLUMEHYD is from a DOS file in batch mode. To create the input file, use any convenient text editor that saves files as ASCII text such as the EDIT program that comes with DOS or the NOTEPAD editor that comes with Windows. Save the input file in any name you wish; however, HYD.IN is the default name

---

* This program is available by downloading it on the Internet in the EPA PLUMES package using the address ftp://ftp.epa.gov/epa_ceam/wwwhtml/software.htm. Select the INSTALPL.EXE program. Then select the area on your computer where you want to install the programs. This will install the PLUMES interface with the EPA UM two-dimensional diffuser model and the RSB line diffuser model along with the PLUMEHYD model.

expected by PLUMEHYD. When you run PLUMEHYD, it prompts you for the input and output file names. **Input** consists of free format metric unit values separated by a space as follows:

Line 1: A one-line title containing anything you want to enter.
Line 2: a. The total number of ports.
         b. The number of diffuser sections.
         c. The dimensionless density difference $(\rho_a - \rho)/\rho$ where ρ_a is the ambient density and ρ is the effluent density.
Line 3: The word "bell" or "sharp" designating the type of entrance to the ports.
A series of lines for each diffuser section starting from the downstream end giving:
         a. The port number for the port at the downstream end of the section.
         b. The port number for the port at the upstream end of the section.
         c. The manifold pipe diameter (m).
         d. The distance between ports (m).
         e. The slope of the manifold given as rise over run starting from the downstream end.
         f. The port diameter (m).
The last line contains Manning's n and the total flow (m³/s).

The input file would look something like the following example:

    Sample input for diffuser
    15 2 0.0267
    sharp
    1 7 1.22 7.3 0.001 .20
    8 15 1.5 7.3 0.0015 .25
    0.014 0.1818

In this example there are 15 ports in two sections. The dimensionless density difference is 0.0267. The first diffuser section starting from downstream has ports 1 through 7 that are 0.2 m in diameter and have a sharp entrance. The ports are in a 1.22-m diameter manifold pipe and are spaced 7.3 m apart. The manifold slope is 0.001. The next section has ports 8 through 15. They are 0.25 m in diameter, spaced 7.3 m apart on a 1.5-m pipe. Manning's n is 0.014 and the total flow is 0.1818 m³/s.

   Output is directed to an ASCII file you specify. It consists of an echo of the input variables for each diffuser section and a line for every port in the section giving the port number, specific energy, E_{j_n}, the discharge coefficient, C_n, the velocity in the pipe, the discharge velocity and flow, and the port Froude number.

Example 2.1
   A diffuser is to discharge 34 mgd (million gallons per day) through 40 4-in. ports spaced 5 ft apart. The diffuser manifold consists of three sections with the

furthest reach having 10 ports and the remaining two having 15 ports each. In order to keep the sediment stirred up, it is desired to have the average velocity be about 3 ft/s in the manifold. If the effluent has a density of 999.16 kg/m³ and the ambient has a density of 1000.1 kg/m³, and the ports are to have sharp entrances, determine the total head loss in the diffuser and the flow distribution. The diffuser is on a sloping bottom of 0.001, and Manning's *n* can be taken as 0.015.

Solution:
      Since the flow distribution is not known, the average flow rate in each can only be estimated. The total flow is 1.5 m³/s when converted from mgd. If we assume an equal distribution to start, the average flow in each section would be 1.22 m³/s, 0.656 m³/s, and 0.188 m³/s. For a 3 ft/s (use 1 m/s) average velocity, this gives pipe diameters of 1.246, 0.914, and 0.484 m, respectively. We now have everything we need to generate the input file.

      For the first line, let's just make the title "Example 2.1 title."
      For the second line:
            a. The total number of ports is 40.
            b. There are three diffuser sections.
            c. The density ratio is (1000.1 − 999.13)/999.13 = 0.001.
      So, the input line is 40 3 0.001
      For the third line we'll enter "sharp."
      For the fourth line:
            a. The first port is number 1.
            b. The last port in this first section is 10.
            c. The manifold diameter is 0.484 m.
            d. The distance between ports is 1.524 m (5 ft).
            e. The slope is 0.001.
            f. The port diameter Is 0.102 (4 in.).
      So, the input line is 1 10 0.484 1.524 0.001 0.102
The fifth line is for the second section from the end with ports 11–25 and becomes

      11 25 .914 1.524 0.001 .102

The sixth line is for the third and last section from the end with ports 26–40
      and becomes

      26 40 1.246 1.524 0.001 .102

The last line is Manning's *n* and the flow, or 0.015 1.5.

      Combining them together, the input file looks like the following:

      Example 3.1 title
      40 3 0.001
      sharp
      1 10 .484 1.524 0.001 .102
      11 25 .914 1.524 0.001 .102
      26 40 1.246 1.524 0.001 .102
      0.015 1.5

If a file named HYD.IN already exists, we either need to delete it or rename it before saving the above data as HYD.IN. When this input file is used when running PLUMEHYD, the output becomes the following:

```
Example 3.1 title
Number of ports      = 40
drho/rho             = 0.0010
Number of sections   = 3
sharp

Mannings N           = 0.0150
Desired Q            = 1.5000
Calculated Q         = 1.5000
```

Friction factor F = 0.0357   Pipe diameter = 0.4840
Length between ports = 1.5240   dz between ports = 0.0010
Port diameter   = 0.1020

Port number	Specific energy (m)	Coeff cd	Pipe velocity (m/s)	Port velocity (m/s)	Port discharge (m^3/s)	Port Froude #
1	2.6621	0.6295	0.2020	4.5490	0.0372	143.8307
2	2.6623	0.6282	0.4036	4.5394	0.0371	143.5271
3	2.6632	0.6259	0.6046	4.5240	0.0370	143.0385
4	2.6653	0.6228	0.8046	4.5032	0.0368	142.3807
5	2.6690	0.6188	1.0034	4.4776	0.0366	141.5708
6	2.6748	0.6141	1.2010	4.4477	0.0363	140.6268
7	2.6831	0.6085	1.3970	4.4142	0.0361	139.5677
8	2.6943	0.6022	1.5914	4.3777	0.0358	138.4131
9	2.7088	0.5953	1.7841	4.3388	0.0355	137.1833
10	2.7270	0.5877	1.9750	4.2982	0.0351	135.8989

Friction factor F = 0.0289   Pipe diameter = 0.9140
Length between ports = 1.5240   dz between ports = 0.0010
Port diameter   = 0.1020

Port number	Specific energy (m)	Coeff cd	Pipe velocity (m/s)	Port velocity (m/s)	Port discharge (m^3/s)	Port Froude #
11	2.8215	0.6261	0.6118	4.6574	0.0381	147.2583
12	2.8224	0.6253	0.6698	4.6524	0.0380	147.0996
13	2.8235	0.6245	0.7276	4.6470	0.0380	146.9294
14	2.8248	0.6235	0.7854	4.6413	0.0379	146.7484
15	2.8263	0.6226	0.8432	4.6353	0.0379	146.5570
16	2.8281	0.6215	0.9008	4.6289	0.0378	146.3558
17	2.8300	0.6204	0.9584	4.6222	0.0378	146.1454
18	2.8323	0.6192	1.0159	4.6153	0.0377	145.9262
19	2.8348	0.6180	1.0733	4.6081	0.0377	145.6989

20	2.8377	0.6167	1.1306	4.6007	0.0376	145.4641
21	2.8408	0.6153	1.1878	4.5931	0.0375	145.2222
22	2.8443	0.6139	1.2449	4.5852	0.0375	144.9739
23	2.8481	0.6124	1.3019	4.5772	0.0374	144.7198
24	2.8522	0.6109	1.3588	4.5690	0.0373	144.4604
25	2.8568	0.6093	1.4156	4.5606	0.0373	144.1963

Friction factor F	= 0.0260	Pipe diameter	= 1.2460
Length between ports	= 1.5240	dz between ports	= 0.0010
Port diameter	= 0.1020		

Port number	Specific energy (m)	Coeff cd	Pipe velocity (m/s)	Port velocity (m/s)	Port discharge (m^3/s)	Port Froude #
26	2.8770	0.6235	0.7931	4.6839	0.0383	148.0955
27	2.8780	0.6230	0.8245	4.6809	0.0382	147.9985
28	2.8791	0.6225	0.8558	4.6777	0.0382	147.8990
29	2.8803	0.6219	0.8871	4.6745	0.0382	147.7972
30	2.8815	0.6213	0.9184	4.6712	0.0382	147.6930
31	2.8829	0.6207	0.9497	4.6678	0.0381	147.5866
32	2.8844	0.6201	0.9810	4.6644	0.0381	147.4781
33	2.8860	0.6195	1.0122	4.6609	0.0381	147.3676
34	2.8876	0.6189	1.0434	4.6573	0.0381	147.2552
35	2.8894	0.6182	1.0746	4.6537	0.0380	147.1410
36	2.8913	0.6175	1.1058	4.6501	0.0380	147.0252
37	2.8932	0.6168	1.1369	4.6464	0.0380	146.9077
38	2.8953	0.6161	1.1680	4.6426	0.0379	146.7888
39	2.8976	0.6153	1.1991	4.6388	0.0379	146.6686
40	2.8999	0.6146	1.2302	4.6350	0.0379	146.5471

Figure 2.7 is a plot of the flow distribution along the length of the diffuser. As can be seen from this figure and the column from the output giving port flow, the flow is fairly uniform. Flows in those ports that vary more than desired can be modified by adjusting the port diameter. Several trials would need to be made to achieve an equal distribution.

The total head loss can be found from the port available head. Since this available head is the difference between the energy grade line in the diffuser to the head outside the port, the energy loss can be determined in this example from the available energy at the 40th port from the end. It is given in the table above as 2.8999 m. This is rather a high loss. It is mostly due to the port velocity head and the sharp entrance to the port giving discharge coefficients of the order of 0.62. If bell-shaped nozzles are used, the total loss would drop to a little over 1 m with nozzle discharge coefficients of the order of 0.96. Change the word "sharp" to "bell" in the previous example and rerun. It makes a big difference at these discharge velocities. An economic analysis would have to be made to determine if the cost of manufacturing the bell-shaped nozzles was more than the savings that would occur with the reduced head loss.

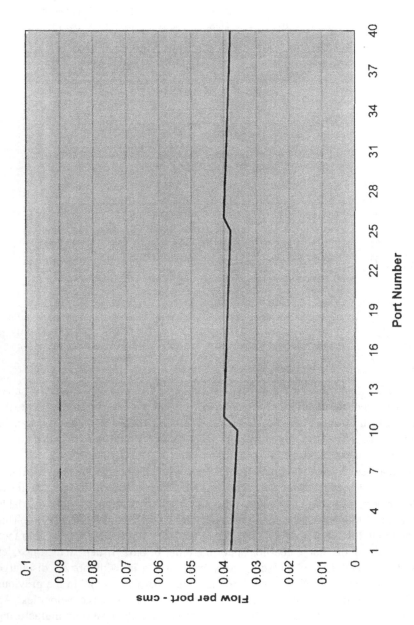

FIGURE 2.7 Flow variation in three-section diffuser in Example 2.1.

Problems

2.1 Repeat Example 2.1 with the average velocity in the diffuser sections limited to 1.0 ft/s. Use 80 3-in. ports in four sections with 20 ports per section. Determine the maximum and minimum flow rates from the ports and the total head loss in the diffuser with both "sharp" and "bell"-shaped nozzles.

2.2 Determine the flow distribution and total head loss in a diffuser designed to discharge 50 cfs from 30 6-in. ports spaced 10 ft apart. There are two sections with 15 ports per section. The average velocity in the diffuser sections is 0.7 m/s. The nozzles have sharp entrances. The effluent density is -0.836 Sigma-t (999.164 kg/m^3) and the ambient density is 26.117 Sigma-t (1026.117 kg/m^3). Use a Manning's n of 0.012 and a bottom slope of 0.002.

2.3 A diffuser is to be designed to discharge 5.0 m^3/s. The minimum average velocity in the diffuser is to be 1.0 m/s, while the average port velocity is to be maintained above 3.0 m/s. The total head loss in the diffuser is limited to 2.0 m. The effluent has the same density as the ambient, and the diffuser can be laid flat on the bottom. Determine the minimum number of diffuser sections that will produce a flow distribution that varies no more than 5%. Use a Manning's n of 0.015.

3 Lagrangian Integral Models

3.1 UM (UMERGE) TWO-DIMENSIONAL SUBMERGED DIFFUSER MODEL

The UM model within the PLUMES package[11] is an updated version of the UMERGE model.[4] It is an integral buoyant free jet model that uses a Lagrangian formulation that considers a cross-sectional slice of the plume or "puff." A Lagrangian formulation has a coordinate system that moves with the plume. Integration is with time. The slice that is followed is usually the shape of a section of a bent cone, as shown in Figure 3.1. Properties within the slice are assumed to be uniform (top hat profiles) but vary along the trajectory with time. The size and shape of the slice vary as a result of plume bending, entrainment, and plume growth. Entrainment of ambient fluid into the slice is divided into two components consisting of forced entrainment due to ambient current through the surface area of the slice seen by the ambient current and Taylor-type aspiration entrainment.

The conservation equations in Lagrangian form used in the model are as follows:

Conservation of mass

$$\frac{dm}{dt} = E_{amb} + E_{\alpha} \tag{3.1}$$

where E_{amb} is the forced entrainment due to ambient velocities, E_{α} is the aspiration entrainment and m is mass.

Conservation of momentum

$$\frac{dm\overline{U}}{dt} = \overline{U}_{\infty}\frac{dm}{dt} - m\overline{g}\frac{(\rho_{\infty}-\rho)}{\rho} \tag{3.2}$$

Conservation of energy

$$\frac{d\left[mC_p\left(T-T_{ref}\right)\right]}{dt} = C_p\left(T_{\infty}-T_{ref}\right)\frac{dm}{dt} \tag{3.3}$$

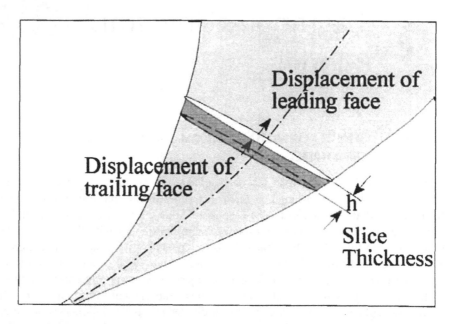

FIGURE 3.1   Plume slice analyzed in the UM Lagrangian model.

Species conservation

$$\frac{dmC_i}{dt} = C_\infty \frac{dm}{dt} - \kappa m C_i \tag{3.4}$$

where C_i are the species of concern, the subscript ∞ refers to ambient conditions, κ is a first-order decay coefficient, T is temperature, ρ is density, $\bar{g}$ is gravity, and C_p is specific heat. The mass within the slice is given by $m = \rho \pi b^2 h$ where b and h are the slice average radius and thickness, respectively. The thickness of the slice can vary with time, since the velocities on each face of the slice are usually different. The density of the slice is normally calculated from the temperature and salinity of the fluid in the slice.

In Equation (3.2), the first term represents the change in total momentum in the slice and is a vector. The second term is the change in momentum caused by the entrainment of ambient fluid. The third term is the change in vertical momentum of the plume due to plume buoyancy caused by the density difference between the plume and ambient. The energy equation assumes that all energy is thermal. No chemical reactions resulting in a change in temperature are considered. In the species equation, the last term allows for the decay of a particular species.

The two entrainment terms in Equation (3.1) are given by

$$E_{amb} = \rho A_p \bar{U} \tag{3.5}$$

where ρ is the ambient fluid density, A_p is the projected surface of the slice that is seen by the approaching ambient fluid, and $\bar{U}$ is the relative vector velocity of the approaching ambient fluid. The second term is

$$E_\alpha = \alpha \rho A_T |\bar{V}| \qquad (3.6)$$

where A_T is the surface area of the slice in contact with the ambient (normally $2\pi bh$), $|\bar{V}|$ is the average plume velocity across the slice, and α is an entrainment function. For "top hat" velocity profiles, this entrainment function can be taken as 0.1.[4] These terms can be evaluated from known starting conditions of a slice. The sum of these terms gives the rate of change of mass in the slice. Once this is determined, the rate of change of momentum, energy, and species concentration can be found from Equations (3.2), (3.3), and (3.4) and known properties in the present slice.

Integration of the above equations in the UM model is performed using a simple single step forward integration such that $m_{t+dt} = m_t + (E_{amb} + E_\alpha)dt$. The new energy, momentum, and species concentration are calculated in a similar manner. This gives new values of

$$m, \quad m|\bar{V}|, \quad mC_p(T - T_{amb}), \quad and \quad mC_i \qquad (3.7)$$

of the slice in its new location. Since the volume of the slice is approximately $\pi b^2 h$,

$$\text{new mass} = \left(\rho \pi b^2 h\right)_{t+dt} = m_{t+dt} \qquad (3.8)$$

and

$$\text{new momentum} = \left(\rho \pi b^2 h |\bar{V}|\right)_{t+dg} = \left(m|\bar{V}|\right)_{t+dg} \qquad (3.9)$$

An integration time step, dt, can be taken such that $dt = h/V,$* which leads to

$$\Delta h = \Delta |\bar{V}| \, dt \qquad (3.10)$$

and the following relation between the new thickness and velocity:

$$h_{t+dt} = h_t \left(\frac{V_{t+dt}}{V_t}\right) \qquad (3.11)$$

---

* This relationship can be expressed is $Vdt = ds$ where ds is the incremental distance along the centerline of the plume. This expression can be used to convert all equations from a time derivative (Lagrangian) to a space derivative (Eulerian), to be discussed in later chapters.

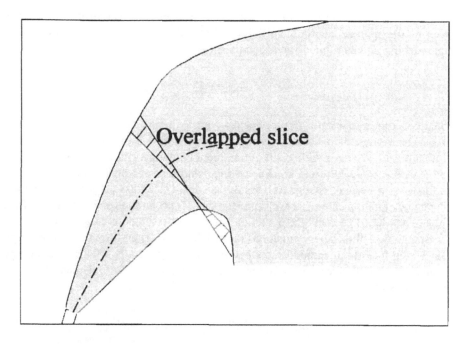

FIGURE 3.2   Plume slice with overlap considered in UM model.

Equations (3.8), (3.9), and (3.11) are three equations with three unknowns, V, b, and h at the new time. Equations (3.3) and (3.4) give the new temperature and concentration. The distance and direction moved are found from the vector velocity and time step. Special considerations are given when plume bending causes the two faces of the slice to overlap, as shown in Figure 3.2 This happens when plumes bend over rapidly.

When plumes from a multiple port discharge merge, the entrainment is modified to reflect the decrease in exposed entrainment surface and the volume of the slice is modified to reflect the symmetry planes between plume centerlines, as shown on Figure 3.3. This reduced area affects both the forced and aspirated entrainment.

The UM formulation is only two-dimensional and considers a plane parallel to the ambient current. As a result, it is recommended that its use be restricted to two-dimensional plumes with the current perpendicular to the diffuser line. Different vertical discharge angles are allowed in that plane but cross-current angles are not considered.

Integration of the equations is performed in a step-wise manner calculating plume trajectory, size, average temperature and concentration, and plume average dilution. Ambient velocity, temperature, and salinity are used at each vertical location the plume passes through. These ambient properties are input in tabular form by the user as a function of depth.

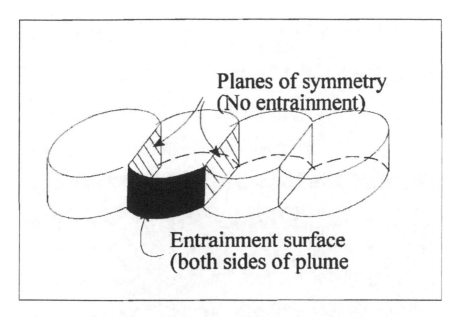

Planes of symmetry (No entrainment)

Entrainment surface (both sides of plume

FIGURE 3.3   Cross section of merging plumes with area considered by UM model.

3.2   UM INPUT AND THE PLUMES INTERFACE

Inputting variables into and running the UM model is best accomplished by means of the EPA PLUMES interface.* It was written by Walter Frick.[11] It is a very convenient program to input and check variables. The interface appears as a spread sheet, as shown on Figure 3.4. The user moves from cell to cell filling in desired information. The program checks for consistency and converts units as required. It calculates dependent variables from values entered for independent variables. For interdependent variables, those entered first become the independent variables and the latter ones become dependent. Independent variables are shown in yellow characters, while dependent variables are shown in white characters. If the user tries to enter values into these latter dependent cells, the program issues a warning and allows the user to decide which are really the independent variables. Pulldown menus and quick keys help the user with the program. The variables are divided into blocks having different colored backgrounds. Outfall variables are magenta, effluent variables are brown, ambient variables are green, and miscellaneous variables are gray. The special red variables are discussed below in the definitions.

Once all cells are filled as desired, the user has several options. The data can be saved as a case, and a new case opened within the same project. The UM or RSB

---

* The PLUMES program and a complete user's manual can be downloaded from the Internet using the address ftp://ftp.epa.gov/epa_ceam/wwwhtml/software.htm.

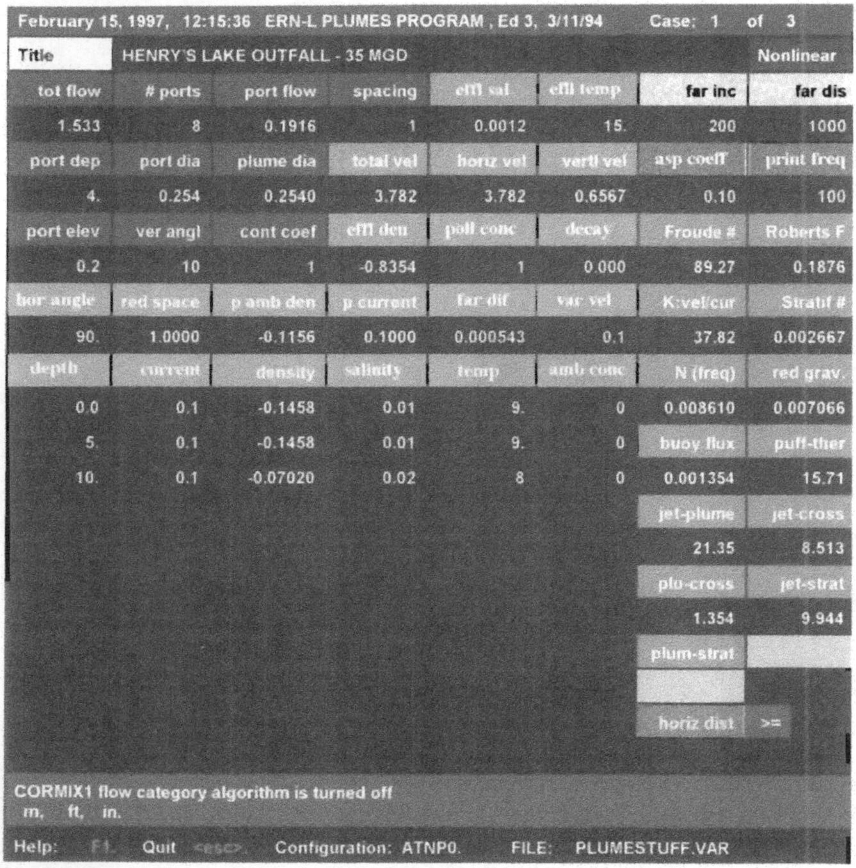

February 15, 1997, 12:15:36 ERN-L PLUMES PROGRAM , Ed 3, 3/11/94						Case: 1	of 3
Title	HENRY'S LAKE OUTFALL - 35 MGD						Nonlinear
tot flow	# ports	port flow	spacing	effl sal	effl temp	far inc	far dis
1.533	8	0.1916	1	0.0012	15.	200	1000
port dep	port dia	plume dia	total vel	horiz vel	vertl vel	asp coeff	print freq
4.	0.254	0.2540	3.782	3.782	0.6567	0.10	100
port elev	ver angl	cont coef	effl den	poll conc	decay	Froude #	Roberts F
0.2	10	1	-0.8354	1	0.000	89.27	0.1876
hor angle	red space	p amb den	p current	far dif	var vel	K:vel/cur	Stratif #
90.	1.0000	-0.1156	0.1000	0.000543	0.1	37.82	0.002667
depth	current	density	salinity	temp	amb conc	N (freq)	red grav.
0.0	0.1	-0.1458	0.01	9.	0	0.008610	0.007066
5.	0.1	-0.1458	0.01	9.	0	buoy flux	puff-ther
10.	0.1	-0.07020	0.02	8	0	0.001354	15.71
						jet-plume	jet-cross
						21.35	8.513
						plu-cross	jet-strat
						1.354	9.944
						plum-strat	
						horiz dist	>=

CORMIX1 flow category algorithm is turned off
m, ft, in.

Help: F1   Quit <esc>   Configuration: ATNP0.   FILE: PLUMESTUFF.VAR

FIGURE 3.4 Typical PLUMES interface screen.

programs can be run directly, or the data can be saved as an ASCII file that can be used by other programs such as the UDKHG program to be discussed later.

3.2.1 PLUMES INTERFACE VARIABLES

Variables in the PLUMES interface are shown in metric units. Output is also in metric units. Variables, however, can be entered in other units by using the ^K option (control K). For example, if the cursor is in the *tot flow* cell, flow is shown in cubic meters per second. Flow can be entered in either cubic meters per second, million gallons per day (mgd), or cubic feet per second (cfs), by successively pressing the ^K keys. This cycles through the options. The values selected are shown in the lower left-hand corner of the sheet. When entered, the program automatically converts the units given into metric units, When the cursor is placed in the *tot flow* cell, the flow in all three units are shown at the bottom-left. Try it. This procedure can be repeated with any input variable that has units. Most common units have been incorporated.

Once entered, a space bar or right arrow key enters the data and moves to the next cell. The other arrow keys can also be used to move around the interface. If values already exist in a cell, the right arrow key only moves one character at a time allowing for the user to edit a portion of the input.

Table 3.1 shows the PLUMES input variables and their definitions.

Now that we have defined all the variables on the PLUMES interface screen, we are ready to discuss the quick keys that can be used to process and execute the program. They are executed by holding down the control key and pressing one or two other keys to select the operation. The F1 function key brings down the main pulldown menu shown on Figure 3.5, which shows the primary quick key commands. Figure 3.6 is the miscellany menu created with the ^Y command. Figure 3.7 is the configuration menu created by the ^R command. Table 3.2 gives all the commands and their operation.

3.2.2   FAR FIELD CALCULATION IN THE PLUMES INTERFACE

When far field calculations are specified in the PLUMES interface, they are calculated by both the constant eddy diffusion model proposed by Brooks[12] and the 4/3 power law model. The results of both are printed out and the user can choose which is felt to be more realistic for the case in concern. The constant eddy diffusion model is a general purpose model that gives approximate far field dilutions in both open coastal and inshore sites. The equation relating pollutant concentration to distance traveled can be expressed as

$$C_{cl} = C_{mz} \operatorname{erf} \sqrt{\frac{U_a b^2}{16 \varepsilon X}} \tag{3.12}$$

where C_{cl} is the pollutant concentration in the center of the waste field at the end of the far field, C_{mz} is the pollutant concentration at the end of the UM calculations, erf is the standard error function, U_a is ambient velocity, b is the width of the waste field including all ports, ε is the lateral eddy diffusivity, and X is the distance traveled from the end of the mixing zone to the end of the far field.

The 4/3 power law is more accurate in open oceans. Dilutions using this method are given by the equation

$$S = \frac{S_a}{\operatorname{erf} \sqrt{\frac{1.5}{\left(1 + 8\alpha b^{4/3} \frac{t}{b^2}\right)^3 - 1}}} \tag{3.13}$$

where S is the centerline dilution in the far field, S_a is the dilution at the end of the UM calculations, t is the time of travel, and α is a dispersion coefficient related to waste field size and eddy diffusivity by $\alpha = \varepsilon/b^{4/3}$. The dispersion coefficient, α in $m^{2/3}/s$, is the term shown in the PLUMES interface. It varies from 0.0001 in low turbulent environments to 0.0005 in higher turbulent environments. Specific values should be used if available. Otherwise the default value is o.k. for approximate answers.

TABLE 3.1
PLUMES Variables

Variable name	Description
Title	A descriptive title as desired by the user. In order to get the cursor to move to the title cell, the ^J (Jump) option must be used. Control J causes the cursor to *jump* from one major section of the interface to another. Try it successively to see where the cursor goes.
tot flow	Total discharge rate from all ports.
# ports	Total number of ports on diffuser.
port flow	Flow from each port. This becomes a dependent variable if tot flow and # ports are entered.
spacing	Spacing between each port. They must all be the same and on one side of the diffuser. Discussions later address the problem of alternating or apposing ports.
effl sal	Salinity in the effluent — parts per thousand.
effl temp	Temperature of the effluent in degrees Celsius.
far inc	If the user elects to calculate far field dilutions using the internal approximate methods, this is the output interval in meters.
far dis	If the user elects to calculate far field dilutions, this is the maximum distance it will make calculations. If the two far field variables are left blank, no far field calculations are carried out.
port dep	This is the depth of the port below the water surface. If the diffuser slopes, the depth of the middle port can be used or several runs can be made using different depths to get a range of dilutions.
port dia	Exit diameter of the ports (all are assumed to be the same).
plume dia	This is the actual diameter of the discharge jet, D_o. It is usually a dependent variable calculated from the port diameter and contraction coefficient given below.
total vel	This is the total jet velocity, U_o. It is interdependent with port flow, plume diameter.
horiz vel	The horizontal component of the total velocity and is dependent on the vertical discharge angle.
vertl vel	The vertical component of the total velocity.
asp coeff	The aspiration coefficient, α, used to calculate part of the entrainment. Unless specific values are known, this value should be unchanged.
print frq	This is how many integration steps are performed between printout. For a value of 500, only a few lines of output result. Many lines of output occur for a value of 10. The value used depends on how much detail is desired. A value of 100 gives output when the dilution is approximately values 2, 4, 8, 16, etc.
port elev	The distance of the port above the bottom. It is used to determine when the plume interacts with the bottom.
ver angle	The vertical discharge angle relative to the horizontal. A value of zero means horizontal discharge in the direction of the current. A value of 90 is for vertical discharge upward. Negative discharge angles are allowed.

TABLE 3.1 (continued)
PLUMES Variables

Variable name	Description
cont coef	The port contraction coefficient; accounts for the vena contracta. A value of 1.0 is typical of a rounded entrance nozzle. A value of 0.6 can be used for a square entrance orifice.
effl den	The density of the effluent given in sigma-t units. Sigma-t is the density in $kg/m^3 - 1000$. A sigma-t of -2 corresponds to a density of 998 kg/m^3. Other units of density can be entered using the ^K command. The density is interdependent with the effluent temperature and salinity. The relationship between density, temperature, and salinity is an empirical equation contained within PLUMES.
poll conc	The initial concentration in the effluent of a particular pollutant of concern. It can be in any units desired.
decay	The decay coefficient, k, in the first-order decay equation $c = c_{max}e^{-kt}$. It can be entered as rate per day, per hour, or as t90hr which is the hours required for the pollutant to decay 90% such that only 10% remains. For conservative pollutants the value of k is zero.
hor angle	This is the horizontal orientation of the diffuser. A value of 90 is for a diffuser manifold pipe which is at 90° (perpendicular) to the ambient current. This is the angle correctly described by the UM model. Values different than 90° for skewed diffusers can be entered for use with other models such as RSB and UDKHG. A value of zero is for diffusers parallel to the current. The RSB model will approximate this case. For a single port, a value of zero gives a port discharging normal to the current such as a pipe coming from the shore out into a river. This case is handled by the UDKHG model.
red space	When a horizontal angle different from 90° is entered, the PLUMES interface calculates the spacing between ports as seen by the ambient current. This is spacing times the sine of the horizontal angle. This spacing is used by the UM model to approximate skewed diffusers. This approach is o.k. far away from the discharge and for vertical discharges, but underpredicts the dilution in the near field for other cases, since it ignores the jet-ambient interaction and causes plume merging to occur too soon. .
p amb den and p current	The ambient density and current at port depth taken from the ambient table.
far dif	The far field diffusion coefficient in $m^{2/3}/s$, which is used if the far field calculations are performed. The default value is appropriate for most cases and should not be changed unless more specific data are available.
far vel	The ambient velocity where far field calculations are to be made. This velocity is used to transport the waste field only and does not affect the rate of dilution.

The following variables are for the ambient. A table is generated giving values of current, density, salinity, temperature, and pollutant concentration at selected depths within the ambient.

TABLE 3.1 (continued)
PLUMES Variables

Variable name	Description
depth	Depth where data are to be entered starting at the surface and going to a depth greater than or equal to the port depth. Up to ten different depths can be entered.
current	The ambient velocity at this depth, U_a.
density	The ambient density at this depth, ρ_a.
salinity	The ambient salinity at this depth.
temp	The ambient temperature at this depth.
amb conc	The concentration of the tracer of concern in the ambient.

The following variables appear to the right of the PLUMES spreadsheet and have a red background. They are variables that result from the input. They are not always of concern to someone doing a mixing zone study, but provide information to modelers on the characteristics of the discharge.

Froude #	The densimetric Froude number of the discharge. It is the ratio of inertia to buoyancy and is given by $Fr = U_o / \sqrt{g D_o (\rho_a - \rho_o)/\rho_a}$ where o and a refer to the discharge and ambient conditions, respectively. A small value between 1 and 5 is for a highly buoyant plume. A value between 5 and 30 or so is for a mixed buoyant jet. Large values are for momentum jets. A value less than one implies ambient intrusion into the diffuser.
Roberts F	A measure of the relative importance of ambient forced entrainment and jet-induced entrainment. For Roberts F less than 0.1, jet-induced (aspirated) entrainment is greater than ambient forced entrainment. For values greater than 0.1, the reverse is true.
K:vel/cur	The ratio of the discharge velocity to the ambient current at discharge depth.
Stratif #	A measure of the ambient stratification given by $(\rho_{ao} - \rho_{as}) D_o/(\rho_{as} - \rho_o) \, depth$, which is the difference in the ambient density at discharge depth and surface times the port diameter divided by the difference between the ambient density at discharge depth and the effluent density times the port depth.
N (freq)	The Brunt–Vaisalla frequency of internal waves. It is given by $$\sqrt{\frac{g(\rho_{ao} - \rho_{as})}{\rho_{ao} \, depth}}.$$
red grav.	The $g\Delta\rho/\rho$ term used in the Froude number.
buoy flux	Buoyancy flux given by $U_a^3 D_o$/Roberts F.
puff-ther, jet-plume, jet-cross, plu-cross, jet-strat, plu-strat	Various transition lengths used in the calculation of the CORMIX category.
hor dis>=	This is the pauSe cell and gives special conditions to stop calculations. The default is horizontal distance. Calculations will stop at this horizontal distance. The variable used in this pause cell can be selected using the ^YS command described below. It is convenient to force output at the edge of a defined mixing zone or at a prescribed dilution. If this variable is left blank, program execution stops when

---

TABLE 3.1 (continued)
PLUMES Variables

Variable name	Description
	the upper edge of the plume reaches the surface or the plume is trapped. An option flag can be set so the plume is allowed to oscillate about the trapping level a given number of times. The cursor can only be placed in this cell by using the ^J (jump) command.

At the bottom, the PLUMES spreadsheet gives the status of configuration, the name of the file being used, and the CORMIX option condition. The current date, the version of PLUMES being used, and which case of the present project is active are displayed at the top. These are discussed later.

When the cursor is placed on any cell, the range of accepted values are given on the lower right hand corner of the PLUMES interface screen.

```
            Main menu
    run rsB program
    run Um program
    show Independents
    units Konversion
    List equations
    get Work file
    fill New file
    add to Output
    cell Precision
    Shallow/surface Z
    configuRe models >
    moVement commands >
    miscellanY menu >
    <esc>
```

FIGURE 3.5   UM pull down main menu obtained with the F1 key showing quick key commands.

3.3 UM OUTPUT

The UM output echos the input values then prints out a table of selected output variables as a function of distance along the plume's trajectory. Figure 3.8 is a sample of UM output that includes far field calculations. The number of lines printed (detail)

```
┌─────────────────────────────────┐
│          Miscellany Menu        │
│      ambient column Fill        │
│      Interpolate amb cell       │
│      Copy ambient line          │
│      Delete ambient line        │
│      Beget new cases            │
│      cHeck consistency          │
│      Notes                      │
│      clear Output cell          │
│      Purge cases                │
│      construct Udf file         │
│      pauSe cell                 │
│      cormiX category            │
│      Zap most variables         │
│      <esc>                      │
└─────────────────────────────────┘
```

FIGURE 3.6   PLUMES miscellany pull down menu obtained by pressing ^Y (Control Y).

```
┌─────────────────────────────────┐
│        Configuration Menu       │
│      Auto ambient               │
│      Brooks eqn input           │
│      Cormix1 categories         │
│      Farfield start             │
│      Reversal set               │
│      Show configuration         │
│      <esc>                      │
└─────────────────────────────────┘
```

FIGURE 3.7   PLUMES configuration menu obtained by pressing ^R (Control R).

is determined by values of the "print frq" and "far inc" cells. The default variables printed out are

a. plume dep, plume centerline depth below the water surface at this point
b. plume dia, plume diameter or width after merging as shown on Figure 3.3
c. poll conc, average pollutant concentration in the plume
d. dilution, average dilution in the plume
e. hor dis, horizontal distance traveled at this point

TABLE 3.2
Quick Key Commands and Their Operation

Keystroke	Operation
^B	This command instructs PLUMES to run the RSB model and then prompts the user which cases to simulate and where the output is to be sent. Output can be sent directly to the screen using "console," to a file by specifying the path and file name, or directly to a connected printer using "prn."
^C	The new case command. It allows the user to select the number of a case within the present project and place it in the active screen or create a new case using the present active screen as a start. Watch the numbers in the upper right-hand corner of the screen when the new case is selected. This command is convenient to group different cases of the same project in one file rather than have a different file for each case.
^U	The command instructs PLUMES to run the UM model with the similar user prompts as with ^B.
^I	Causes the variables that are interdependent with the variable where the cursor lies to be shown with a black hatched background.
^K	Selects the input units to be entered for the cell where the cursor lies and converts the value entered to metric units.
^L	Gives the definition of the selected cell along with the equations used in its evaluation. This command is very helpful if you are not sure what goes in the cell. Try moving around to different cells and execute the ^L command to see what you get.
^W	Is used to select another working file of records or cases. When executed, a window appears where the name of the file can be selected. Those files ending in .VAR can be cycled through using the ↓ (down arrow) key. When selected, the existing active file is saved and the new data is brought into PLUMES. The default data file is PLMSTUFF.VAR. If a new file name is typed in, the existing active file is saved and the screen is cleared of all but default data.
^N	Creates a new file of records from the active records. A new screen appears and the user is prompted for the new file name. A name with the extension .VAR is suggested so it can be recognized by the ^W command. A second prompt appears asking which cases are to be saved on the new records. This is possible since the present active set of variables may consist of several cases as described below. If several cases are selected, separate them with a space. Sequential cases can be separated by two periods such as 2..5 (two through five)
^O	This directs the interface to add the present variable to the printed output. Output variables are indicated by having a blue square in its background field.
^P	Changes the precision of dependent variables with a maximum of six significant figures.
^Z	Tells PLUMES to use the image method to approximate single port discharge into shallow water. This is done by using image plumes to force lines of symmetry which simulate contact with the surface and bottom of the receiving water. This is discussed in more detail using UDKHG.
^R	Brings up the configuration menu.
^V	Brings up the cursor movement menu. These commands are similar to the commands used in some word processors. They are not used too often since the arrow, delete, and backspace keys on most PCs do the same thing.
^Y	Brings up the miscellany menu.

TABLE 3.2 (continued)
Quick Key Commands and Their Operation

Keystroke	Operation
^RA	This is a two-stroke command accomplished by holding down the control key, pressing R then A. It brings up the configuration menu and executes option A in one process. ^RA toggles auto-filling in the ambient table on and off. If on, PLUMES automatically fills in the value of the independent variable above it into the cell when filling in the ambient table. This helps if different depths have some of the same fluid properties. The status of the setting is given as the first character in the configuration string such as ATNO0 (on) or NTNO0 (off).
^RB	Toggles the option to use with the Brooks far field model. It is given as the second character in the configuration string such as NRCO0 or NTCO0. The R option allows the user to set the waste field width and origin to start the far field calculations. With the T option, these values are Transmitted from the UM calculations.
^RC	This toggles the third character of the configuration string between C and N. With the C option, PLUMES will calculate the CORMIX flow class. With N it will not.
^RF	This toggles the fourth character of the configuration string between M, O, and P. With M, far field calculations are initiated when the plume reaches its maximum height of rise. With the O option, far field calculations begin when the faces of the plume "slice" overlap causing uncertain predictions. With the P (pause) option, the conditions set in the Pause cell, ^YP, are used.
^RR	When a plume reaches neutral buoyancy in a stratified environment, it will overshoot, reach a maximum, and return back toward the neutral buoyancy level, overshooting in the opposite direction. These oscillations (and Reversals) about the neutral buoyancy level will continue until all vertical momentum has been dissipated. This option sets how many of these reversals are allowed before execution stops or far field calculations are initiated. The default is zero (0) which stops UM at the first reversal for positive buoyancy and the second for negative plumes. Negative plumes are often discharged upward. They reach a maximum then drop. If the ambient is stratified, it could reach a neutral buoyancy level and bend back up again. This is taken as the default stopping point for negative plumes.
^RS	Brings up a window showing the status of the configuration.
^YF	This causes blank cells in the column of the ambient table to be filled with the value given in the surface cell.
^YI	This command causes PLUMES to fill an ambient column, Interpolating linearly with depth between values that are given.
^YD	Deletes an ambient line
^YC	Copies a line in the ambient block and places it between it and the next ambient line.
^YB	Allows the user to copy the contents of the selected cell to the same cell in other selected cases. When executed, a window is opened prompting the user as to which cases the value is to be copied.
^YH	Check for inconsistencies in input values as compared with accepted limits given by PLUMES. Try some weird value in a cell and execute ^YH and see what it tells you.
^YN	Displays the last 20 messages that have appeared in the dialog window at the bottom.
^YO	Cells that already are in the output list can be removed using the ^YO command.
^YU	An important command that writes the contents of the PLUMES interface into an ASCII text file called UDF.IN that can be used as input to the UDKHG submerged diffuser

---

TABLE 3.2 (continued)
Quick Key Commands and Their Operation

Keystroke	Operation
	program. If data already exists in the UDF.IN file, the data specified will be appended to that already in the file. In this way, multiple cases can be placed in one file to be run by UDKHG.
	It can also read data from an existing UDF.IN file into the interface. When executed, the user is prompted whether he wants to read a file into the interface or write to the UDF.IN file. If the write option is selected, the user is prompted for which cases are to be written. If the read option is selected, the user is prompted as to whether the file is to be added as a new case to the present project or to replace the present case.
^YS	This is used to set up and edit the pauSe cell (below the red block of variables). For example, the program can be told to stop when the centerline concentration goes below a specified value. The ^YS command brings up a window or dialog box that allows the user to select the pause variable, the type of inequality or whether the variable is to be added to the output list. The Pause option in the configuration must be selected for this pause cell to be effective. Execute ^YS and then cycle through the variables using V to see what variables can be set in the pause cell.
^YX	Calculates the CORMIX flow class (see ^RC)
^YZ	This command clears all cells in the active screen and fills some with the default values ready for a new case.
<esc>	The escape key cancels the present operation or quits the PLUMES interface.

Other output columns can be generated using the Output option of the pause cell, such as centerline concentration, plume temperature, time of travel, or density difference. Output for the far field calculations consists of the waste field average concentration and dilution as predicted by both far field models, the distance in increments selected in the PLUMES interface, and the time of travel in seconds and hours.

Notice that the UM model flags the points where the bottom edge of the plume encounters the bottom, plumes merge, and when the upper edge of the plume hits the surface. Calculation stopped in this case when the plume hit the surface. Note that only two elevations were given in the ambient table. The UM model linearly interpolates between ambient entries to get specific velocities and densities at intermediate depths during calculations.

3.4 WINDOWS INTERFACE FOR SIMULATING PLUMES (WISP)

A Windows interface that uses an updated version of the UM model to predict plumes has been recently developed by Dr. Walter Frick of the EPA. The major difference is in the input procedure and analysis capability. Rather than using the PLUMES interface that runs through DOS, WISP uses a WINDOWS interface that allows you

```
Sep  2, 1996,  12:14:57  ERL-N PROGRAM PLUMES, Dec 21, 1992   Case:   3 of   4
Title   Oldport Outfall 8 10° dia ports  - 1 meter apart
tot flow   # ports port flow   spacing  effl sal effl temp   far inc   far dis
   1.533        8   0.1916         1   0.0012     15.         200      1000
port dep   port dia plume dia total vel horiz vel vertl vel asp coeff print frq
   4.        0.254   0.2540      3.782    3.724    0.6567      0.10       100
port elev ver angle cont coef  effl den poll conc     decay  Froude #  Roberts F
   0.2        10        1     -0.8354        1      0.000    89.27     0.1876
hor angle red space p amb den p current   far dif   far vel K:vel/cur Stratif #
  90.     1.0000   -0.1156    0.1000  0.000453     0.1     37.82    0.002667
  depth   current  density  salinity      temp   amb conc  N (freq)  red grav.
  0.0        0.1   -0.1458     0.01       9.        0    0.008610  0.007066
 10.         0.1   -0.07020    0.02       8         0  buoy flux  puff-ther
                                                       0.001354     15.71
                                                      jet-plume  jet-cross
                                                         21.35     8.513
                                                      plu-cross  jet-strat
                                                          1.354     9.944
                                                      plu-strat
                                                          6.787
                                                      hor dis>=

CORMIX1 flow category algorithm is turned off.
  m,  ft,  in.                                                    to  m range
Help: F1.   Quit: <esc>.  Configuration:ATNP0.  FILE: PLMSTUFF.VAR;
 plume dep plume dia poll conc  dilution   hor dis
      m         m                              m


   4.000      0.2540     1.000      1.000     0.000
   3.892      0.4997     0.5000     1.999     0.6198
   3.687      0.9750     0.2500     3.998     1.840
   3.673      1.008      0.2415     4.139     1.926< merging
   3.481      1.458      0.1830     5.461     3.106< bottom hit
   2.942      2.752      0.1250     7.995     6.507< bottom hit
   2.202      4.519      0.09408   10.62     11.23< surface hit< bottom hit
Farfield calculations based on Brooks (1960), see guide for details:
Farfield dispersion based on wastefield width of      11.52m
  --4/3 Power Law--    -Const Eddy Diff-
     conc  dilution       conc  dilution distance        Time
                                                  m       sec    hrs
   0.03627      27.6    0.05760      17.4    200.0    1888    0.5
   0.01782      56.1    0.04260      23.5    400.0    3888    1.1
   0.01103      90.7    0.03530      28.3    600.0    5888    1.6
   0.007668    130.4    0.03080      32.5    800.0    7888    2.2
   0.005725    174.7    0.02767      36.1   1000.0    9988    2.7
```

FIGURE 3.8   PLUMES interface screen with example 3.1 values.

to use the mouse to select cells to be filled. It also separates the ambient files from the discharge data. In the discharge screen, you can enter several values for each of the variables such as discharge rate, port diameter, and port spacing as well as selecting several ambient files to be considered. You can then run a particular case, run cases sequentially, or have the program run all combinations of input variables. This saves time in running a matrix of runs. It also has the capability of running a simple time series analysis.

If you are interested in getting WISP and the manual that goes with it, contact Dr. Frick directly. His e-mail is *WALTER.FRICK@epamail.epa.gov.*

3.5 EXAMPLES

Example 3.1

A diffuser is to be placed in a coastal environment and discharge 40 mgd of treated municipal waste. The controlling pollutant is chlorine. The concentration of chlorine

```
Sep  2, 1996,  15:13:59  ERL-N PROGRAM PLUMES, Dec 21, 1992   Case:   1 of   1
Title    Example 3.1 - 40 mgd - site No. 1
 tot flow    # ports port flow    spacing effl sal effl temp   far inc    far dis
    1.753         30  0.05843      3.048     0.1       15
 port dep  port dia plume dia total vel horiz vel vertl vel asp coeff print frq
  30.48      0.1524    0.1524      3.203    3.010     1.096     0.10        100
 port elev ver angle cont coef  effl den poll conc     decay   Froude # Roberts F
    1          20       1.0      -0.7592     0.5         0       16.36 0.0002936
 hor angle red space p amb den p current   far dif   far vel K:vel/cur Stratif #
    90        3.048    24.87     0.03048  0.000453               105.1 0.0003494
 depth     current   density   salinity      temp  amb conc  N (freq)  red grav.
   0.0      0.06096    23.08       30         10         0     0.02371   0.2516
  10       0.06096    23.55       30.5        9.5        0     0.01470   4.276
  15       0.06096    24.02       31          9          0  jet-plume jet-cross
  20       0.04572    24.41       31.5        9          0     2.347      14.19
  30       0.03048    24.87       32          8.5        0  plu-cross jet-strat
  35       0.03048    24.87       32          8.5        0     519.1      4.271
                                                            plu-strat
                                                               5.762
                                                            dilution>=
                                                               45.5

CORMIX1 flow category algorithm is turned off.
0.06096 m/s, 0.2000 ft/s                                  >=0.0 to 2.0 m/s range
Help: F1.   Quit: <esc>.   Configuration:ATNP0.   FILE: TEST.VAR;
```

FIGURE 3.9   PLUMES interface screen with example 3.1 values.

in the effluent is 0.5 ppt (parts per thousand, o/oo). The required concentration of chlorine at the end of the mixing zone is 0.011 ppt. The resulting target dilution is 0.5/0.011 = 45.5. The effluent temperature and salinity are 15°C and 0.1 ppt, respectively. A preliminary design has 30 6-in. ports placed 10 ft apart on one side of the diffuser manifold. The manifold pipe is to be placed in 100 ft of water, perpendicular to the prevailing current and the ports are to discharge up 20° from the horizontal. The nozzles will be manufactured with rounded approaches. A survey of the site location yields the following ambient conditions:

Depth (m)	Velocity (ft/s)	Salinity (o/oo)	Temperature (°C)
0	0.20	30.0	10.0
10	0.20	30.5	9.5
15	0.20	31.0	9.5
20	0.15	31.5	9.0
30	0.10	32.0	8.5
35	0.10	32.0	8.5

The problem is to determine whether this design will cause the chlorine content within the plume to reach 0.011 before it reaches the surface or is trapped. Save the data for the PLUMES interface in a file called test.var.

Solution:

When the PLUMES interface screen is filled in with appropriate values for this problem, it looks like Figure 3.9. The ^K command was used to simplify the input so that the English units could be entered directly where necessary. The ^J command was used to place the cursor in the title and in the pause cell. The ^YS command

plume dep	plume dia	poll conc	dilution	hor dis
m	m			m
30.48	0.1524	0.5000	1.000	0.000
30.35	0.2998	0.2500	1.975	0.3476
30.08	0.5873	0.1250	3.925	1.032
29.44	1.112	0.06250	7.825	2.316
27.78	1.915	0.03125	15.62	4.378
24.27	3.016	0.01563	31.23	6.880
24.09	3.071	0.01520	32.10	6.983< merging
20.86	4.162	0.0106	45.71	8.644

FIGURE 3.10   UM output for example 3.1.

and the Variables option were used to select dilution as the pause variable and the ^RF command was used so turn on the Pause cell. This sets the configuration to ATNP0. A cutoff value of 45.5 was entered in the pause cell to stop execution when the target dilution is reached. Save the sheet as test.var by pressing ^N and entering the name TEST.VAR in the space provided. When prompted for which cases, enter a 1.

An investigation of the sheet shows a Froude number of 16.4 and a small Roberts F. This means the discharge has both momentum and buoyancy and that entrainment is mainly by aspiration. Stratification is weak so the plume may not be trapped. After you have adjusted the input variables to look like Figure 3.9, run the UM model by executing ^U and hit the space bar twice. Since no output file has been specified, the present case is run and the output is sent to the screen. The output section of the screen should look like Figure 3.10. Since no far field calculations were specified, the program stopped when the condition of the pause cell was met. In this case, UM stopped on the next step after the dilution reached 45.5. This occurred when the plume was still submerged at a depth of 20.86 m and 8.64 m downstream. Thus, the target dilution was reached before the plume reached the surface or trapped.

Out of curiosity, turn the pause cell off and rerun the case just to see what happens to the plume. This is done by using the ^RF command and selecting Max-rise or Overlap. Use the ^YS command to change the pause variable to hor-dis. When you rerun the UM model again, it continues to make calculations until the upper edge of the plume hits the surface. At this point the plume thickness (diameter) is 26.8 m, which nearly fills the water column. The dilution at this point is 114.9. You notice, however, that the plumes have merged and passed through the neutral buoyant point. It was starting to bend over when the upper edge hit the surface. When it started to bend, overlap of adjacent slices occurred and UM issues a message indicating that calculations beyond that point are questionable.

Example 3.2

Several alternate diffusers and flows are to be considered for the discharge in Example 3.1. Set up the cases in the same project file and determine if the required dilution is obtained or not. The cases to be considered are:

Case 1.   The case given in Example 3.1

Case 2.   The same conditions given in Example 1 except there are ten 12-in. ports spaced 10 ft apart in 30 ft of water. The stratification table remains the same.

Case 3.   The same case as 2 except a higher discharge rate of 60 mgd is to be considered in 40 ft of water.

Case 4.   The same case as 1 except the 30 ports are to be placed on alternate sides of the diffuser to make it shorter and the discharge depth is to be 30 ft.

Solution:

Since most variables are the same as case 1, the file created in Example 3.1 can be used as a template. To generate an input file for case 2, go to the input file for Example 3.1. If you have left the PLUMES interface and have not worked on another project, just execute the PLUMES interface again. It automatically brings up the last file you worked on. If you have changed things and want the original file, press ^W and use the down arrow to select TEST.VAR from the file window at the bottom. It should be there since we saved it in Example 3.1. Press enter when you have it. Create a new case by pressing ^C. PLUMES will automatically increment the case by one (in this case it will show a 2 in the prompt screen). Now change the values on this screen according to case 2 conditions by changing the number of ports to ten, diameter to 12 in., and depth to 30 ft. Be sure to change the title to reflect the case.

Now create case 3 by pressing ^C again. It is automatically numbered 3 of 3. Now change the flow 60 mgd, the depth to 40 ft and title for this case. Since case 4 is a variation of case 1, go back to case 1 by pressing ^C and entering 1 in the prompt window. Case 1 of 3 appears. Now type ^C again and enter a 4 in the prompt window. Now you have 4 of 4 with the data of case 1.

The fact that case 4 has a diffuser with alternating ports requires that some simplifications be made, as discussed in Chapter 2. We can assume that the two sides are independent of each other and consider half the flow from 15 ports (optimistic approach) or assume all 30 ports are on one side of the diffuser with only 5 ft spacings (conservative approach). Let's take the optimistic approach for case 4 and create a case 5 for the conservative approach. The actual dilution will be in between the two approaches. Change the depth to 30 ft, title, the flow to 20 mgd, and number of ports 15 to set up case 4. Create case 5 with a ^C command and change its title, flow to 40 mgd, number of ports to 30, and spacing to 5 ft to set up case 5.

We can now run each case separately by using the ^C command to select the case wanted and then press ^U. Press the space bar to opt for just one case and select the file name for the output. We can also opt to run all cases at once and place them in one output file by going to case 2, pressing ^U and when PLUMES prompts for the number of cases, enter a 4 (that's cases 2, 3, 4, and 5). Then give the name of the file where you want the output.

Figures 3.11 to 3.14 are the output sections for each case as generated by UM. As can be seen, all plumes surfaced and only the optimistic approach to the alternating diffuser (case 4) gave a required dilution of 45.5. Since the actual dilution

plume dep	plume dia	poll conc	dilution	hor dis	
m	m			m	
9.144	0.3048	0.5000	1.000	0.000	
8.882	0.5924	0.2500	1.976	0.6902	
8.275	1.122	0.1250	3.929	1.999	
6.698	1.952	0.06250	7.834	4.187	
3.393	3.064	0.03169	15.43	6.875<	merging
3.301	3.092	0.03125	15.65	6.934	
0.9050	3.842	0.02368	20.64	8.336<	surface hit

FIGURE 3.11   UM output for example 3.2, case 2.

plume dep	plume dia	poll conc	dilution	hor dis	
m	m			m	
12.19	0.3048	0.5000	1.000	0.000	
11.93	0.5971	0.2500	1.976	0.6943	
11.38	1.157	0.1250	3.928	2.048	
10.04	2.139	0.06250	7.833	4.525	
8.051	3.069	0.03901	12.54	6.992<	merging
6.308	3.688	0.03125	15.64	8.606	
1.347	5.282	0.02134	22.89	12.05<	surface hit

FIGURE 3.12   UM output for example 3.2, case 3.

plume dep	plume dia	poll conc	dilution	hor dis	
m	m			m	
9.144	0.1524	0.5000	1.000	0.000	
9.017	0.2987	0.2500	1.976	0.3477	
8.754	0.5809	0.1250	3.929	1.031	
8.156	1.091	0.06250	7.834	2.321	
6.663	1.889	0.03125	15.64	4.470	
3.803	3.046	0.01563	31.27	7.088	
3.731	3.075	0.01541	31.70	7.144<	merging
1.115	4.242	0.01075	45.45	8.993<	surface hit

FIGURE 3.13   UM output for example 3.2, case 4.

plume dep	plume dia	poll conc	dilution	hor dis	
m	m			m	
9.144	0.1524	0.5000	1.000	0.000	
9.017	0.2987	0.2500	1.976	0.3477	
8.754	0.5809	0.1250	3.929	1.031	
8.156	1.091	0.06250	7.834	2.321	
7.416	1.536	0.04124	11.86	3.516<	merging
6.420	1.998	0.03125	15.64	4.735	
1.024	4.058	0.01584	30.84	8.871<	surface hit

FIGURE 3.14   UM output for example 3.2, case 5.

will probably be less than that due to plume interaction with the other side, this is not a good diffuser either. By taking the conservative approach of case 5 and trying several runs with the diffuser at different depths, it is found that by placing the diffuser 55 ft deep, a dilution of 45.5 could be achieved with confidence. This is a cheaper solution to the problem than case 1.

Problems

3.1 A landfill has 7,000,000 gallons of leachate that it must dispose of each year. The leachate has a chloride (Total Dissolved Solids) content of 17,000 mg/l with a specific gravity of 1.03. Regulations for the freshwater river where the leachate is to be discharged limit the chloride concentration to 100 mg/l with a 100 ft mixing zone. Because of low river flow, disposal is not allowed during the nine lowest river flow months. This leaves only about 100 days of the year when discharge to the river is allowed. During these months the lowest river flow is 200 ft³/s. The depth in the center of the river during this flow is 8 ft, which gives an average velocity of 1.2 ft/s. Is there enough river flow to provide the required dilution during the 100 days? Can a diffuser be designed to achieved the required dilution withing the mixing zone?

3.2 A 16-port line diffuser is to be used to discharge 0.26 m³/s of waste water into a 5-m deep river. The ports are 10 cm in diameter spaced 3 m apart on one side of the diffuser. Discharge is horizontal, and the diffuser is perpendicular to a 12 cm/s current. The density of the effluent and ambient are essentially the same. If the effluent is initially 10° warmer than the ambient, what is the average plume temperature 5 m downstream?

3.3 81.1 mgd of treated municipal waste are discharged through a 25-port diffuser into a coastal zone where the average depth is 15 m. The ports are spaced 10 m apart and are oriented in the direction of the prevailing current and up 5° from the horizontal. Port diameters of 6, 10, and 12 in. are to be considered. The problem is to find the dilution at the end of the zone of initial dilution where the plume is trapped or hits the surface when

a. The ambient is unstratified with a specific gravity of 1.025 and zero current.

b. When the ambient density and current vary as given by the following table:

Depth — m	Density gm/cm³	Velocity — m/s
0	1.0184	0
3	1.0204	0
6	1.0205	0
9	1.0207	0
12	1.0208	0
16	1.0208	0

c. The same as b above but with a 0.12 m/s current.

4 Eulerian Integral Methods

4.1 UDKHG 3-D SUBMERGED DIFFUSER MODEL

Eulerian methods look at a fixed control volume and the changes in properties of the fluids that pass in and out of the volume. Equations are expressed in space derivatives rather than with time as in Lagrangian methods. The UDKHG* is a version of UKDHDEN[4,13] that has the option of displaying the output graphically on the computer screen. It uses a Eulerian approach and integral methods to convert the governing partial differential equations to a series of ordinary differential equations. This method has been found to produce very good approximations to fluid behavior as long as boundary effects do not enter the problem.[14,15] Integral methods require that appropriate distribution profiles be assumed for velocity, temperature, and concentration. To accommodate merging of multiple plumes, the UDKHG program uses 3/2 power law profiles as explained in Chapter 1.

4.1.1 THEORETICAL DEVELOPMENT

The basic computational scheme of UDKHG is based on the analysis of Hirst.[16,17] He developed a three-dimensional analysis of a single port discharge into a flowing stratified environment, either air or water. He developed the axisymmetric equations of motion and energy from the Navier–Stokes and energy transport equations. Figure 4.1 is a defining sketch showing a general three-dimensional plume, and the coordinates used. By assuming symmetry and integrating in the radial direction, these equations were reduced to the following form:

Conservation of mass

$$\frac{d}{ds} \int_0^\infty \bar{u}\, r\, dr = E \qquad (4.1)$$

This equation says that the change in flow rate within the plume is equal to the fluid entrained, E. This entrained fluid is due to forced entrainment, aspirated entrainment, and buoyant thermal-type entrainment and is described below.

---

* The UDKHG program can be downloaded from the World Wide Web using the address www.engr.orst.edu/~ldavis.

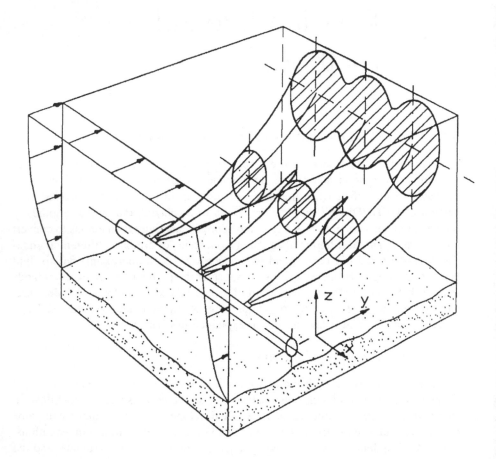

FIGURE 4.1   Three-dimensional plume with coordinate definitions.

Conservation of energy

$$\frac{d}{ds}\int_0^\infty \bar{u}\left(\bar{T} - \bar{T}_a\right)r\,dr = -\frac{d\bar{T}_a}{ds}\int_0^\infty \bar{u}\,r\,dr - \lim_{r\Rightarrow\infty}\left(\overline{rv'T'}\right) \qquad (4.2)$$

The first term of this equation represents the total excess thermal energy in the plume. The second term represents the change in excess thermal energy due to temperature stratification (the ambient reference is changing). The last term with the turbulence term v' is the exchange of thermal energy due to ambient turbulence at the edge of the plume. Heat capacity is assumed constant and cancels from all terms. The species conservation equation is the same as the energy equation except concentration replaces temperature.

Conservation of momentum along the trajectory

$$\frac{d}{ds}\int_0^\infty \bar{u}^2 r\, dr = \bar{U}_a E \sin\theta_2 + \int_0^\infty g(\rho - \rho_0) r\, dr \sin\theta_2 - \lim_{r\Rightarrow\infty}(\overline{ru'v'}) \qquad (4.3)$$

The first term of this equation is the change in momentum in the centerline direction of the plume. The second term is the change in momentum due to ambient entrainment. The third term is the change in momentum due to the density difference between the plume and ambient, and the last term is the ambient turbulence contribution to plume momentum changes (Reynolds stresses at edge of the plume). The angles give the proper components of the s momentum. The angle θ_2 is the plume flow direction relative to the horizontal and θ_1 is the plume flow direction relative to the current. The momentum equation can be divided into components to relate changes in flow direction. These are given as the curvature equations as follows:

$$\frac{d\theta_2}{ds} = \frac{g\int_0^\infty (\rho - \rho_\infty) r\, dr \cos\theta_2 - EU_a \sin\theta_1 \sin\theta_2}{\bar{q}} \qquad (4.4)$$

and

$$\frac{d\theta_1}{ds} = \frac{EU_a \cos\theta_1}{\bar{q}\cos\theta_2} \qquad (4.5)$$

where

$$\bar{q} = \int_0^\infty \bar{u}^2 r\, dr - \frac{E^2}{4} - \lim_{r\Rightarrow\infty}(r^2\overline{v'^2}) \qquad (4.6)$$

The accuracy of the calculations depends to a large extent on how precise the entrainment is approximated. Kannberg and Davis[13] used the following expression for the entrainment:

$$E = \left(a_1 + \frac{a_2}{F_L}\right)\left[b|u_c - U_a \cos\theta_2|\left(1 - \frac{a_4 b}{L}\right) + a_3 U_a \sin\theta_2\right] \qquad (4.7)$$

the coefficients, where a_i are experimentally determined coefficients and F_L is a local Froude number. The first term on the right represents the Taylor-type aspirated entrainment and the buoyant thermal-type entrainment. The second term on the right gives the decrease in fluid entrainment due to competition between adjacent plumes for the same fluid and is a function of how big the plume is, b, and the spacing between plumes, L. The last term is the forced entrainment due to ambient current.

In the zone of flow establishment, the profiles change from the assumed "top hat" shape at the end of the discharge nozzles to bell-shaped profiles. Following the zone of flow establishment is the region of fully developed single plumes, where the profiles remain similar, changing only in magnitude. In the zone of merging plumes, these single plumes slowly merge into one another. This merging region can occur anywhere along the trajectory of the plume, depending upon the ambient conditions and the initial port spacing.

The zone of flow establishment in the UDKHG program is calculated by applying the above equations to free boundary layer profiles with a central core that decays as the shear layer grows. The length of the development zone is determined when the central core disappears. A development zone in the UM model is not calculated since profiles are assumed to be "top hat" the whole way. As a result, the development zone is ignored.

It has been found by experiment that jet turbulence dominates entrainment in the near field and ambient turbulence has little effect. As a result, the ambient turbulence terms in the governing equations are usually ignored for mixing zone calculations.[16] With these terms ignored, the governing Equations (4.1)–(4.7) can be integrated when the velocity, concentration, and temperature profiles are selected. In UDKHG, 3/2 power law profiles are assumed similar to Equation (1.3) for velocity, temperature, salinity, and species concentration. The integrated equations become a system of ordinary differential equations in terms of centerline values and plume width with the streamwise coordinate, s, as the independent variable. These equations are solved in the UDKHG program in a stepwise manner along the plume centerline using Hamming's predictor–corrector method with a fourth-order Runge–Kutta method giving starting values. The program automatically checks for accuracy by checking for the sum of the changes in all properties in each integration step. If changes are too big, the integration step, ds, is successively halved until the sum of all changes is within accuracy limits. The results of the program produce the three-dimensional trajectory of the plume, the plume size, centerline excess temperature, species concentration, the excess velocity in the plume, and the average dilution.

The plumes begin to merge when the width of the individual plumes becomes equal to the spacing between the ports. This has to be taken into account in the calculations, since the profiles are no longer axisymmetric once the plumes have started to merge. In addition, the entrainment surface is less than the circumference because of the merged region. Figure 4.2 shows a cross section of merging plumes, and Figure 4.3 shows the approximate shape of the concentration distribution along the line connecting the center of each individual plume with the overlap region where plumes have merged into each other. In order to continue the integral analysis through the merging zone, the profiles in the merging plumes must be known. Superposition of overlapping distributions for single plumes furnishes the approximation as shown on Figure 4.3. The temperature profile in the r direction in a line connecting plumes in this merging region is given by

$$\Delta T_r = \Delta T_c \left\{ \left[1 - \left(\frac{r}{b} \right)^{3/2} \right]^2 + \left[1 - \left(\frac{L-r}{b} \right)^{3/2} \right]^2 \right\} \tag{4.8}$$

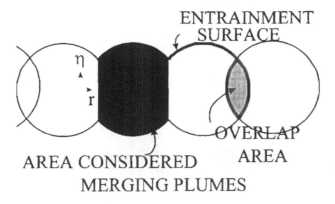

FIGURE 4.2   Cross section of merging plumes showing overlap regions.

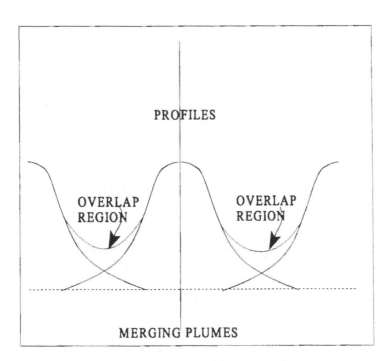

FIGURE 4.3   Concentration distribution along connecting line of merging plumes with overlap regions.

In the η direction perpendicular to the r direction, it is given by

$$\Delta T_\eta = \Delta T_r \left[1 - \left(\frac{\eta}{c} \right)^{3/2} \right]^2 ; \quad c = \sqrt{b^2 - r^2} \tag{4.9}$$

```
Sample input for UDKHG: 5-1 meter ports - 20 m3/s discharge
 0 0 0 0 0 0 0 0    1 0 1 0          10         2.
        20.          5    1.00       15.      70.      0.10        1.
        0.400       60.    20.       0.0      0.0    0.0003        0.0
          5          0.    20.    0.99827     1.        0.  0
        0.0         28.    17.       0.21   1.02019     0.
        20.         29.    12.       0.213  1.02197     0.
        40.         30.     8.       0.2135 1.02338     0.
        60.         30.     5.       0.214  1.02375     0.
        80.         30.     5.       0.214  1.02375     0.
```

FIGURE 4.4   Sample of a UDKHG input file as generated by PLUMES and saved as a UDF.IN file.

Similar expressions are used for velocity, density, and concentration. When entered into the governing equations, they can be integrated to give ordinary differential equations that apply to the merging region. When completely merged, properties are assumed to only vary in the η direction. The entrainment surface is only that in contact with the ambient, as shown in Figure 4.2, and varies with the degree of merging.

4.1.2   UDKHG INPUT

The easiest way to generate the input file for the UDKHG program is to use the EPA PLUMES interface discussed in Chapter 3. Once all of the input cells are filled, the input file for UDKHG can be created using the Save Universal Data File (UDF.IN) option, ^YU, in the PLUMES interface. Multiple cases can be saved on the same UDF file by saving another case without deleting the old one. Figure 4.4 is an example of the UDF.IN file. The name of the UDF.IN file should be changed if the file is to be kept.

Usually, the user does not need to know the format of the UDF.IN file: he just uses it as generated by PLUMES. In some cases where the file consists of a series of runs that the user wants to edit without going back to PLUMES, the appropriate variables can be changed in the UDF.IN file using a standard ASCII text editor. In that case, the format of the file needs to be known. It is as given below with variables in metric units with all lengths in meters, temperature in degrees Celsius, and salinity in o/oo.

Line one: Title
Line two: Option flags. Only the first zero applies to UDKHG. If it is changed to a one, UDKHG makes this case interactive. It will prompt the user for a title for that case and then prompt the user what changes are to be made for the next case after the present case is run. All the other flags in this line are ignored by UDKHG. Try it and see what happens.
Line three: Total flow, number of ports, discharge diameter, vertical angle, port depth, and two more variables not used in UDKHG.

```
THE DISPLAY SETTINGS ARE:
1. Display mode:
   X-Y plane and Z-Y plane are displayed
2. Display Area [m]:
   Min_X: -40; Min_Y: -40;  Min_Z:  -40
   Max_X: 70: Max_Y: 240; Max_Z:  50
3. Gradation of Concentration Distribution
        Concentration Ratio   Color
              0.6               4
              0.2               3
              0.10              5
              0.05             13
4. Display legend on right
5. Distance from bottom to ports

If you wish to change settings, enter number.
Enter "D" for display of results  >
```

FIGURE 4.5   Graphic interface menu used in UDKHG allowing the graphic output to be configured by the user.

Line four: Ambient velocity at surface, horizontal angle, spacing between ports, and four variables not used in UDKHG.

Line five: Number of lines in ambient table, discharge salinity, discharge temperature, discharge density in grams per cubic centimeter, and three variables not used in UDKHG.

Ambient table with each line containing depth, salinity, temperature, ambient current, density, and a variable not used in UDKHG.

4.1.3 UDKHG OUTPUT

The program is written in Fortran and compiled and linked with a standard compiler having a graphic library. When UDKHG is executed, the user is prompted for the name of the input file to be used and where the output should be sent. After the program runs a case, the user is prompted to choose (a) just graphics output, (b) just tabulated output on file, or (c) both. Figure 4.5 is the graphics interface menu obtained when either the display or both options are selected. The graphical display shows the plume, its trajectory, and the concentration distribution within the plume

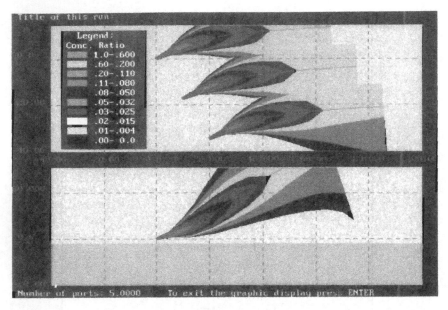

FIGURE 4.7 Typical UDKHG graphic output showing both plan and elevation views of a diffuser plume. Colored bands represent concentration bands with ranges given in legend.

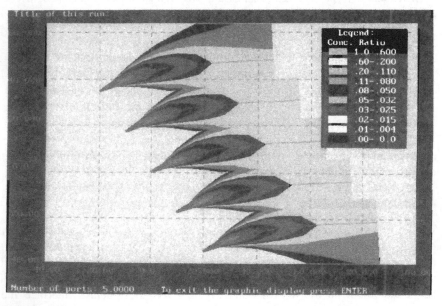

FIGURE 4.8 Replot of Figure 4.7 with just the plan view selected from the interface menu and a smaller region of interest zoomed in on.

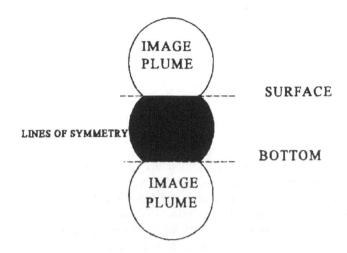

FIGURE 4.9   Image plumes simulating single port discharge into shallow water.

4.1.4   Using UDKHG in Shallow Water

At many sites, especially in inland waters, the ambient is shallow. This causes the plume to rapidly interact with the bottom and surface, impeding entrainment from those sides. UDKHG can be used to approximate such discharges provided the following conditions are met: the discharge densimetric Froude number is greater than 5, the vertical discharge angle is less than 30, discharge is from a single port or from ports wide enough apart that they do not merge in the region of interest, and the ambient depth is greater than about twice the discharge port diameter.[19]

The procedure is to use UDKHG's merging routine with the method of images used in mathematics. Since the ambient is shallow, the vertical trajectory is ignored and the plume is assumed to be centered in the water column. The SPACING between the ports variable is set to the water depth and the DEPTH variable is set to a dummy large number. This forces the image plumes to simulate the top and bottom, as shown in Figure 4.9. Experiments have found this method to give good results as long as the conditions given above are not violated.[19,20] In the output, the Z coordinate is ignored and the merging statement indicates where the plume fills the water column. The user must check the plume width against the port spacing to ensure that the plumes do not actually merge while simulating the top and bottom with image plumes. Obviously, the plume can't grow bigger than the river or entrain more water than is available. The user must check this manually.

For multiple port discharges in shallow water, nothing else has to be done other than that already discussed. For single port discharges, however, some games have to be played to force the image method. This is done by telling the program there is more than one port, then multiplying the flow rate by the number of ports entered. This ensures that the central port, which UDKHG simulates, has the right flow rate. Telling the program there is more than one port causes it to execute the merging routine when the plume size grows to the water depth (which you entered earlier as spacing).*

4.2   UDKHG EXAMPLES

Example 4.1

Solve Example 3.1 using UDKHG. Compare the results of the two programs.

Solution:

You could have the data for Example 3.1 saved either as a .VAR file or as an UDF.IN file. If you saved it in a UDF.IN file, you are ready to run UDKHG. If you saved the data using the ^N command in PLUMES, it can be recalled using the ^W command, using the down arrow to select the .VAR file you saved it under. Once the correct data are in the PLUMES interface, save the file using the ^YU command, which saves the data in the UDF.IN file. If you didn't save the data for Example 3.1, enter it again into the PLUMES interface and save it using the ^YU command. If you already have a UDF.IN file with data in it, you may want to rename or delete it before saving the data of Example 3.1; otherwise, you will have multiple runs in the same UDF.IN file.

Now run the UDKHG program by going to the directory where it was saved when downloading from the Web and typing UDKHG. When prompted for the name of the input file, enter the path and UDF.IN (or the name you changed it to). When prompted for an output file name, enter a nonexisting file name or type TTY or CON to just send the output to your screen. In this example, let's send it to a file called EXM4.1 and elect to just send the output to the file (don't select the graphic output option). After the program runs, view the results using a standard text editor such as MS-WORD, WordPerfect, EDIT in DOS, or the Clipboard in Windows.** Figure 4.10 is the output. Notice that the required dilution of 45.5 occurs about 7 m downstream where the plume is 6.5 m above the discharge. This is not quite the same as was found using the UM program. This is to be expected, since turbulent computational fluid dynamics is not an exact science and two different schemes are used. They are close, however, and either could be used. (More on accuracy will be discussed later.)

Example 4.2

Repeat Example 4.1 using the graphic output to determine the results.

Solution:

Rerun Example 3.1 data as in Example 4.1, except select the option to have the output go to the screen. When the graphic interface screen comes up, enter D to

---

* It should be noted that UM through the PLUMES interface also has a simplified version of the image method for shallow water discharges. It is limited to single ports discharging directly downstream. To execute this option in UM, press ^Z and answer the prompt to give the river width.

** Some word processors enter text files using a default font that has proportional spacing. This causes misalignment of columns in UDKHG output. You can fix this by selecting the whole file (^A in WORD and WordPerfect) and changing the font to Courier 8 point.

```
                            PROGRAM UDKHG - V:1.0
                  SOLUTION TO MULTIPLE BUOYANT DISCHARGE PROBLEM WITH
                  AMBIENT CURRENTS AND VERTICAL GRADIENTS.   MARCH 1995

        UDKH_GRA Version 1.0 (March 95)
        UNIVERSAL DATA FILE: exm4_1.in
        CASE I.D. Example 3.1 - 40 mgd - site No. 1
        DISCHARGE= 1.7530CU-M/S  * TEMPERATURE= 15.00-C * SALINITY= 0.100-PPT
        DIA-M= 0.1520 * NO. OF PORTS= 30 * SPACING=   3.05-M * DEPTH=  30.48-M
        VERTICAL ANGLE = 20.00 * HORIZONTAL ANGLE = 90.00

               AMBIENT STRATIFICATION PROFILE
            DEPTH (M)        TEMP (C)        SALINITY (PPT)  DENSITY (G/CM3) VELOCITY (M/S)
               0.00           10.00            30.00            1.02308           0.061
              10.00            9.50            30.50            1.02355           0.061
              15.00            9.00            31.00            1.02402           0.061
              20.00            9.00            31.50            1.02441           0.045
              30.00            8.50            32.00            1.02487           0.030
              35.00            8.50            32.00            1.02487           0.030

        FROUDE NO= 16.47,  PORT SPACING/PORT DIA=     20.05
        STARTING LENGTH=    0.905

        ALL LENGTHS ARE IN METERS. FIRST LINE ARE INITIAL CONDITIONS.
           X        Y        Z      DIAM.     DRHO      DTCL      DSCL    DILUTION

          0.00     0.00     0.00     0.15     1.000     1.000     1.000     1.00
          0.00     0.85     0.32     0.42     0.989     0.989     0.989     1.98
          0.00     1.97     0.80     1.30     0.306     0.301     0.304     6.63
          0.00     3.02     1.41     2.14     0.173     0.164     0.172    11.91
          0.00     3.97     2.16     2.97     0.112     0.099     0.112    18.22

        PLUMES MERGING

          0.00     4.83     3.02     3.70     0.079     0.060     0.079    24.59
          0.00     5.60     3.96     4.31     0.059     0.034     0.059    30.51
          0.00     6.31     4.95     4.93     0.043     0.014     0.044    36.39
          0.00     6.97     5.97     5.59     0.031    -0.002     0.032    42.22
          0.00     7.61     7.01     6.27     0.023    -0.014     0.024    47.97
          0.00     8.23     8.06     7.41     0.016    -0.023     0.018    53.71
          0.00     8.84     9.11     8.36     0.012    -0.032     0.014    59.48
          0.00    10.08    11.19    10.57     0.004    -0.043     0.006    71.20

        PLUMES HAVE REACHED EQUILIBRIUM HEIGHT - STRATIFIED ENVIRONMENT

          0.00    11.17    12.85    12.82    -0.004    -0.039    -0.003    80.96
          0.00    11.90    13.82    14.58    -0.008    -0.037    -0.007    86.98
          0.00    12.71    14.72    16.63    -0.013    -0.036    -0.012    92.76
          0.00    13.63    15.51    18.99    -0.017    -0.037    -0.016    97.95
          0.00    14.69    16.10    21.58    -0.021    -0.055    -0.019   101.89
          0.00    15.88    16.34    23.09    -0.023    -0.063    -0.021   103.61

        PLUMES HAVE REACHED MAXIMUM HEIGHT - STRATIFIED ENVIRONMENT

        TRAPPING LEVEL= 18.51 METERS BELOW SURFACE,          DILUTION=  75.72
```

FIGURE 4.10   Tabular output for Example 4.1.

display the results using the default settings. Figure 4.11 shows the results of the plan view. Since the discharge concentration is 0.5 ppt and the required concentration is 0.011 ppt, the desired concentration ratio is 0.011/0.5 = 0.022. From the graphic output screen, it is seen that this occurs 9 and 13 m downstream. Remember that this is centerline concentration, and the dilutions given in Examples 3.1 and 4.1 are averaged across the plume. An average concentration ratio of 0.022 would occur with a centerline concentration of about 0.044. Figure 4.11 shows this to occur between 6 and 8 m downstream, agreeing with the tabulated results.

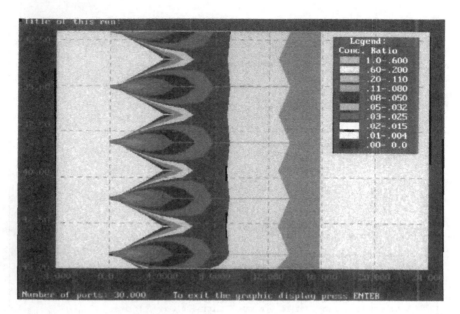

FIGURE 4.11   Graphic output for Example 4.2.

Example 4.3

Repeat Example 3.1 using UDKHG but change the horizontal angle to 60°, which is for a diffuser whose manifold pipe is at 60° (120°) from the current and discharge from the ports is 30° from downstream. Select both file and graphic output and compare the results with Example 4.1

Solution:

Either execute the PLUMES interface with the Example 3.1 data, change the horizontal angle to 60° and save as a UDF.IN file, or edit the file used in Example 4.1 by changing the horizontal angle from 90° to 60°. Rerun UDKHG using the new file and select an output file name of exm4.3. Figure 4.12 is a copy of the output file. From the output file, you will notice that the desired dilution of 45.5 occurs about 6 m downstream, which is closer to the discharge ports than predicted in Example 4.1 and 4.2. This is due to the current–plume interaction (forced entrainment) near the discharge. This effect cannot be measured by the UM model due to its two-dimensionality. At higher ambient to discharge velocity ratios, however, the plumes are rapidly swept downstream and merge quicker due to the relative spacing between ports. The actual dilution for these higher current cases will probably be between UM's more conservative approach and UDKHG's 3-D approach.

Example 4.4

Industrial waste is discharge from a single 12-in. diameter pipe at a rate of 8 ft³/s. The discharge pipe is perpendicular to the shore of a large estuary. At the tidal condition in question, the discharge port is 40 ft below the surface and discharges up 30° from the horizontal. The discharge and ambient stratification are as shown

```
                        PROGRAM UDKHG - V:1.0
               SOLUTION TO MULTIPLE BUOYANT DISCHARGE PROBLEM WITH
               AMBIENT CURRENTS AND VERTICAL GRADIENTS.   MARCH  1995

    UDKH_GRA Version 1.0 (March 95)
    UNIVERSAL DATA FILE: exm4_3.in
    CASE I.D. Example 4.3 - 40 mgd - site No. 1
    DISCHARGE= 1.7530CU-M/S  * TEMPERATURE= 15.00-C * SALINITY= 0.100-PPT
    DIA-M= 0.1520 * NO. OF PORTS= 30 * SPACING=  3.05-M * DEPTH= 30.48-M
    VERTICAL ANGLE =  20.00 * HORIZONTAL ANGLE =  60.00

            AMBIENT STRATIFICATION PROFILE
       DEPTH (M)        TEMP (C)       SALINITY (PPT)   DENSITY (G/CM3) VELOCITY (M/S)
         0.00           10.00            30.00           1.02308          0.061
        10.00            9.50            30.50           1.02355          0.061
        15.00            9.00            31.00           1.02402          0.061
        20.00            9.00            31.50           1.02441          0.045
        30.00            8.50            32.00           1.02487          0.030
        35.00            8.50            32.00           1.02487          0.030

    FROUDE NO= 16.47.  PORT SPACING/PORT DIA=     20.05
    STARTING LENGTH=    0.888

    ALL LENGTHS ARE IN METERS. FIRST LINE ARE INITIAL CONDITIONS.
       X        Y        Z       DIAM.    DRHO      DTCL      DSCL    DILUTION

     0.00     0.00     0.00     0.15     1.000     1.000     1.000      1.00
     0.42     0.72     0.31     0.42     0.990     0.990     0.990      1.98
     0.96     1.70     0.80     1.35     0.295     0.290     0.293      6.85
     1.44     2.63     1.41     2.25     0.164     0.155     0.163     12.48

    PLUMES MERGING

     1.86     3.47     2.18     3.14     0.105     0.092     0.105     19.24
     2.20     4.24     3.06     3.86     0.075     0.056     0.075     25.53
     2.48     4.94     4.01     4.47     0.056     0.030     0.056     31.51
     2.72     5.58     5.02     5.10     0.041     0.011     0.042     37.44
     2.92     6.19     6.05     5.80     0.029    -0.004     0.030     43.31
     3.10     6.79     7.09     6.62     0.020    -0.014     0.021     49.10
     3.27     7.37     8.15     7.68     0.015    -0.025     0.017     54.86
     3.42     7.95     9.21     8.66     0.011    -0.034     0.013     60.71
     3.69     9.14    11.31    10.97     0.002    -0.041     0.004     72.55

    PLUMES HAVE REACHED EQUILIBRIUM HEIGHT - STRATIFIED ENVIRONMENT

     3.86    10.02    12.72    12.96    -0.004    -0.038    -0.003     80.90
     3.99    10.73    13.70    14.77    -0.008    -0.036    -0.007     86.98
     4.12    11.52    14.62    16.92    -0.013    -0.035    -0.012     92.83
     4.26    12.42    15.42    19.40    -0.017    -0.035    -0.016     98.11
     4.41    13.46    16.01    22.13    -0.021    -0.050    -0.020    102.16
     4.57    14.64    16.23    23.73    -0.023    -0.057    -0.021    104.15

    PLUMES HAVE REACHED MAXIMUM HEIGHT - STRATIFIED ENVIRONMENT

    TRAPPING LEVEL= 18.64 METERS BELOW SURFACE,          DILUTION= 75.62
```

FIGURE 4.12   Tabular output for Example 4.3.

on the PLUMES interface, Figure 4.13. Using UDKHG, determine if a dilution of 30 can be achieved before the plume surfaces or is trapped.

Solution:

Delete or rename your old UDF.IN file. Set up the PLUMES interface as shown in Figure 4.13. Save the data as an UDF.IN file. Note the fairly strong variation in ambient density with depth and a zero horizontal discharge angle. The PLUMES-UM model will give poor predictions for this strongly three-dimensional type of discharge. Run the UDKHG program using the "both" option for output. You can

```
Mar 24, 1997,  14:59: 3  ERL-N PROGRAM PLUMES, Ed 3, 3/11/94  Case:    1 of   1
Title    Example 4.4 - Single port - 8 cfs                          nonlinear
 tot flow   # ports port flow   spacing effl sal effl temp   far inc  far dis
    0.2265         1   0.2265      1000      0.1       10
port dep  port dia plume dia total vel horiz vel vertl vel asp coeff print frq
   12.19    0.3048    0.3048     3.104    2.688    1.552      0.10        50
port elev ver angle cont coef  effl den poll conc    decay  Froude # Roberts F
       1        30         1 -0.156774        1        0     11.53
hor angle red space p amb den p current   far dif  far vel K:vel/cur Stratif #
       0     0.000   24.0963    0.1524    0.0003             20.37  0.001046
   depth   current   density  salinity      temp  amb conc N (freq) red grav.
     0.0    0.1524   23.0818        25        10        0   0.02823    0.2379
      10    0.1524   23.5500      30.5       9.5        0 buoy flux puff-ther
      15    0.1524   24.7971        32         9        0               3.919
      20    0.1463   25.5779        33         9        0 jet-plume jet-cross
```

FIGURE 4.13   PLUMES interface screen values for Example 4.4.

use the graphic output to see the interesting shape of the plume. It has an "S" shaped
trajectory with the plume traveling out into the ambient about as fast as it rises due
to buoyancy.

From the output file (Figure 4.14), we see that the plume is trapped 8.1 m below
the surface, about 4.5 m out into the estuary, where the dilution is 36.4. You will
notice that the plume continues to rise past the equilibrium height for about 2 m due
to its upward momentum. It then bends over where the program stops. The dilution
at the maximum height of rise is greater than the value printed out at the point where
it passes through the equilibrium height. In some cases this higher dilution is
acceptable.

Example 4.5

Repeat the discharge of Example 4.4 into a shallow river. The ambient average
depth is 2 m. It has an average velocity, temperature, and salinity of 0.5 m/s, 10°C,
and 0.01 o/oo. The river is 50 m wide with a total flow of 1700 ft³/s. Use the shallow
water image method to determine the average plume dilution 150 ft downstream.

Solution:

Rename or delete the old UDF.IN file. Enter the PLUMES interface with the
data from Example 4.4. Make a new case using the ^C command. You now have
two of two. Change the title, the spacing to 2 m (the depth), and the depth to 50
(the river width). Change the number of ports to 3 and multiply the actual flow by
three (24 ft³/s). Change the ambient table giving just two depths, one at the surface
and one at 50 m. Enter the current, temperature, and salinity of the river on both
lines. The pollutant concentration in the ambient should be set to zero. Save the file
using the ^YU command.

Run the UDKHG program and choose the file only output (graphics will have
no meaning, since there is only one port and we have told the program there are
three to simulate images). Figure 4.15 gives a portion of the output. Notice that
merging (plume fills the water column) occurs about 1 m from the discharge. At
150 ft downstream (Y = 45.7 m), the plume fills the water column, is 8.5 m wide

```
                        PROGRAM UDKHG - V:1.0
               SOLUTION TO MULTIPLE BUOYANT DISCHARGE PROBLEM WITH
               AMBIENT CURRENTS AND VERTICAL GRADIENTS.   MARCH 1995

       UDKH_GRA Version 1.0 (March 95)
       UNIVERSAL DATA FILE: examp4_4.in
       CASE I.D. Example 4.4 - Single port - 8 cfs


            SINGLE PORT DISCHARGE CASE
       DISCHARGE= 0.2265CU-M/S  * TEMPERATURE= 10.00-C * SALINITY= 0.100-PPT
       DIA-M= 0.3040 * NO. OF PORTS=   1 * SPACING=1000.00-M * DEPTH=  12.19-M
       VERTICAL ANGLE = 30.00 * HORIZONTAL ANGLE =    0.00

              AMBIENT STRATIFICATION PROFILE
          DEPTH (M)        TEMP (C)      SALINITY (PPT)   DENSITY (G/CM3) VELOCITY (M/S)
              0.00            9.95           25.00          1.01921          0.152
             10.00            9.50           30.50          1.02355          0.152
             15.00            9.00           32.00          1.02480          0.152
             20.00            9.00           33.00          1.02558          0.146

       FROUDE NO= 11.61,  PORT SPACING/PORT DIA=   3289.47
       STARTING LENGTH=    1.310

       ALL LENGTHS ARE IN METERS. FIRST LINE ARE INITIAL CONDITIONS.
           X       Y       Z     DIAM.    DRHO     DTCL     DSCL    DILUTION

          0.00    0.00    0.00    0.30    1.000    1.000    1.000     1.00
          1.15    0.03    0.69    0.84    0.930    0.845    0.931     1.98
          2.99    0.64    2.10    4.20    0.148   -0.092    0.149    13.29
          4.21    2.11    3.59    6.85    0.027   -0.243    0.028    29.99

       PLUMES HAVE REACHED EQUILIBRIUM HEIGHT - STRATIFIED ENVIRONMENT

          5.13    4.30    4.93    9.15   -0.042   -0.392   -0.041    48.84
          5.80    6.57    5.40   10.45   -0.066   -0.451   -0.064    60.25

       PLUMES HAVE REACHED MAXIMUM HEIGHT - STRATIFIED ENVIRONMENT

       TRAPPING LEVEL=   8.10 METERS BELOW SURFACE.           DILUTION=  36.43
```

FIGURE 4.14 Tabular output for Example 4.4 — Single port discharge, normal to a strongly stratified ambient.

and has an average dilution of 38.5. The maximum possible dilution when the plume is completely mixed is $(1700 + 8)/8 = 213.5$. Since the predicted dilution and plume width are less than the limitations, the results are reasonable.

4.3 INTEGRAL METHODS WITH SURFACE PLUMES — PDSG

4.3.1 INTRODUCTION

Surface plumes are quite different from submerged plumes because of the effect of the free surface. They occur whenever discharge of a buoyant fluid is near the surface or when a side channel discharges into a river, lake, or coastal water. If the discharge is less dense than the ambient, it will be forced up by the more dense ambient. Since the surface acts as a lid, this upward force causes the plume to become thin and

```
                        PROGRAM UDKH_GRA - V:1.0
                SOLUTION TO MULTIPLE BUOYANT DISCHARGE PROBLEM WITH
                AMBIENT CURRENTS AND VERTICAL GRADIENTS.   MARCH 1995

        UDKH_GRA Version 1.0 (March 95)
        UNIVERSAL DATA FILE: exam4_5.in
        CASE I.D. Example 4.5 - Single port - 8 cfs into a shallow river

        DISCHARGE= 0.6796CU-M/S  * TEMPERATURE= 10.00-C * SALINITY= 0.100-PPT
        DIA-M= 0.3040 * NO. OF PORTS=  3 * SPACING=   2.00-M * DEPTH=  15.24-M
        VERTICAL ANGLE =   30.00 * HORIZONTAL ANGLE =   0.00

                AMBIENT STRATIFICATION PROFILE
           DEPTH (M)        TEMP (C)        SALINITY (PPT)   DENSITY (G/CM3) VELOCITY (M/S)
               0.00            9.00            0.10             0.99992         0.500
              50.00            9.00            0.10             0.99992         0.500

        FROUDE NO=200.03,   PORT SPACING/PORT DIA=     6.58
        STARTING LENGTH=     0.978

        ALL LENGTHS ARE IN METERS. FIRST LINE ARE INITIAL CONDITIONS.
             X        Y        Z       DIAM.     DRHO      DTCL      DSCL    DILUTION

            0.00     0.00     0.00     0.30     1.000     1.000     1.000      1.00
            0.84     0.07     0.49     0.83     1.000     1.000     0.000      2.05

        PLUMES MERGING

            2.02     1.90     1.17     2.91     0.211     0.226     0.000     12.91
            2.44     4.28     1.41     3.82     0.127     0.137     0.000     17.21
            2.73     6.69     1.57     4.41     0.101     0.109     0.000     19.94
            2.95     9.11     1.70     4.89     0.091     0.099     0.000     22.14
              .        .        .
            4.01    28.53     2.33     7.25     0.062     0.067     0.000     32.74
            4.18    33.39     2.44     7.66     0.059     0.064     0.000     34.59
            4.35    38.25     2.55     8.04     0.056     0.061     0.000     36.27
            4.50    43.11     2.64     8.39     0.054     0.058     0.000     37.83
            4.63    47.97     2.73     8.71     0.052     0.056     0.000     39.28
            4.76    52.83     2.82     9.02     0.050     0.054     0.000     40.64
```

FIGURE 4.15 Sections of the tabular output from UDKHG for Example 4.5 using the image method with the note "merging" indicating where the single plume fills the water column.

spread out laterally in all directions, resulting in buoyant spreading and reduced vertical entrainment. The resulting plume can be thin and wide with little aspirated entrainment except near the discharge. In addition, if the plume is heated, surface heat transfer reduces plume temperature and plume thermal energy is not conserved as it is in a submerged plume. All this must be accounted for in surface models.

4.3.2 THEORETICAL DEVELOPMENT

The PDSG* model is an integral such as UDKHG and UM where the equations of motion are integrated across the plume assuming representative profiles for velocity, species concentration, and temperature. The theoretical analysis used to develop the three-dimensional surface plume program (PDS) is based on an earlier model by

---

* The PDSG program can be downloaded from the Internet using the address www.engr.orst.edu/~davisl/ plumes.html.

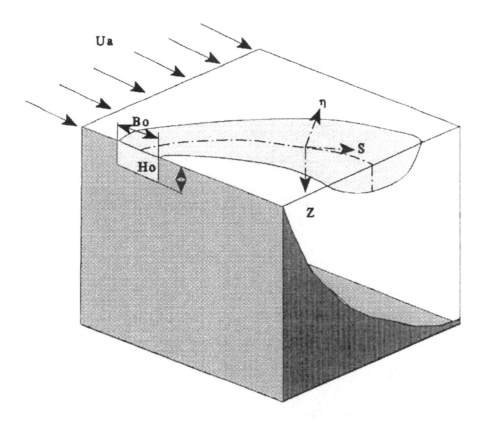

FIGURE 4.16 Sketch of surface plume described by PDSG with local coordinates.

Prych.[21] It was modified to its present form by Davis and was tuned to data by Shirazi, hence PDS.[22] The version of PDS with graphical output, PDSG, will be discussed here and was first presented by Davis and Hoevekamp.[23] It was originally developed for thermal discharges but applies equally as well to other types. It assumes discharge is from one bank into a large receiving body where no bank interaction occurs, as shown in Figure 4.16. The discharge channel is assumed to be rectangular with a depth of H_o and a half width of B_o. The local coordinate system has a Z coordinate going down from the surface and a horizontal η coordinate normal to the plume centerline, as shown in the figure. The profiles assumed are Gaussian in the horizontal direction and half Gaussian in the vertical such that

$$T_r = T \exp\left(-\eta^2 / B^2\right) \cdot \exp\left(-Z^2 / H^2\right) \qquad (4.11)$$

$$U_r = U \exp\left(-\eta^2 / B^2\right) \cdot \exp\left(-Z^2 / H^2\right) + V \cos\theta \qquad (4.12)$$

where T and U are the centerline temperature and velocity respectively, and V is the ambient velocity. The limits on η are from $-\infty$ to $+\infty$. The limits on Z are from 0 to ∞

(down). H and B are the characteristic depth and half width, respectively, defined as $\sqrt{2}$ of the standard deviation of the Gaussian profiles in the vertical and horizontal directions.

With the temperature and velocity profiles assumed, the energy, volume, and momentum fluxes can be integrated across the plume at any cross section, leaving them in terms of centerline values and plume characteristics width, B, and depth, H. Accordingly, the volume flux becomes

$$Q = \iint_A (U_r) d\eta dZ = \pi HB \left(\frac{U}{2} + V \cos \theta \right) \tag{4.13}$$

where the limits of integration for $V \cos \theta$ are taken as the bottom half of the region:

$$\left(\frac{\eta}{\sqrt{2}B} \right)^2 + \left(\frac{Z}{\sqrt{2}H} \right)^2 \leq 1 \tag{4.14}$$

Solving Equation (4.13) for U yields

$$U = 2 \left(\frac{Q}{\pi HB} - V \cos \theta \right) \tag{4.15}$$

The heat flux, J, is

$$J = \iint_A U_r T_r d\eta dZ = \frac{\pi}{2} TBH \left(\frac{U}{2} + V \cos \theta \right) = \frac{QT}{2} \tag{4.16}$$

The momentum flux, M, is

$$M = \iint_A U_r^2 d\eta dZ = \pi BH \left(\frac{U}{2} + V \cos \theta \right)^2 = \frac{Q^2}{\pi BH} \tag{4.17}$$

The changes in flow, temperature, and momentum as given by dQ/ds, dT/ds, and dM/ds are calculated from conservation equations. dQ/ds is assumed to be due to contributions of jet entrainment and ambient turbulent mixing; thus,

$$\frac{dQ}{ds} = \frac{dQ}{ds}\Big|_j + \frac{dQ}{ds}\Big|_a \tag{4.18}$$

The jet and ambient contributions are both divided into vertical and horizontal components. The horizontal jet entrainment fluid is

$$\frac{dQ}{ds}\Big|_{j,h} = 2 E_0 \int_{-\sqrt{2}H}^{0} \Delta U dZ \tag{4.19}$$

where

$$\Delta U = \left(U^2 + V^2 \sin^2 \theta\right)^{1/2} \exp\left(-Z^2/H^2\right)$$ (4.20)

and E_o is an entrainment coefficient. By inserting Equation (4.20) into Equation (4.19), we obtain

$$\left.\frac{dQ}{ds}\right|_{j,h} = \sqrt{\pi} H E_o \left(U^2 + V^2 \sin \theta\right)^{1/2}$$ (4.21)

The vertical jet entrained fluid is

$$\left.\frac{dQ}{ds}\right|_{j,v} = 2 \int_0^{\sqrt{2}B} E\Delta U_v d\eta$$ (4.22)

where $E = E_o f(R_i)$ and R_i is the local Richardson number given by

$$R_i = \frac{\sqrt{2}}{F_o^2} \frac{HT(s,\eta,o)}{\Delta U_v^2}$$ (4.23)

The function $f(R_i)$ is a curve fit to data. It is

$$f = \left[\exp\left(-5 R_i\right) - 0.0183\right]/0.982$$ (4.24)

The velocity difference ΔU_v is given by

$$\Delta U_v = \left[U^2 \exp\left(-2\eta^2/B^2\right) + V^2 \sin^2 \theta\right]^{1/2}$$ (4.25)

The term T is the surface excess temperature at a distance η from the plume centerline. The value of the integral Equation (4.22) is determined numerically in the program.

The effective entrainment due to ambient turbulent mixing is calculated as follows:

$$\left.\frac{dQ}{ds}\right|_{a,h} = 11.0 \frac{H}{B} \frac{\varepsilon_h}{U_o H_o}$$ (4.26)

$$\left.\frac{dQ}{ds}\right|_{a,v} = 11.0 \frac{B}{H} \frac{\varepsilon_v}{U_o H_o} f(R_i)$$ (4.27)

where ε_h and ε_v are the horizontal and vertical turbulent diffusion coefficients, respectively.

The change in heat flux along the plume due to heat exchange with the atmosphere is expressed as

$$\frac{dJ}{ds} = -2 \int_{0}^{\sqrt{2}B} K\,T_r d\eta = \sqrt{\pi KTB} \qquad (4.28)$$

where K is a dimensionless heat exchange coefficient. Substituting Equation (4.28) into Equation (4.20) yields

$$\frac{dT}{ds} = -\frac{T}{Q}\left(2\sqrt{\pi}KB + \frac{dQ}{ds}\right) \qquad (4.29)$$

The conservation of momentum is applied in the s-direction and then divided into X and Y components. The net forces on the plume are balanced by the change in momentum flux. The forces considered important are (a) the internal pressure forces due to buoyancy, and (b) form drag due to ambient current and interfacial shear forces.

The pressure forces are found by determining the excess pressure due to buoyancy as a function of depth and then integrating the pressure over the vertical cross section of the plume. Thus, the normalized pressure force is

$$P = \frac{1}{F_o^2} \iint_A \left(\int_{-\infty}^{z} T_r dZ\right) dA = \frac{\sqrt{\pi}TH^2 B}{2F_o^2} \qquad (4.30)$$

where F_o is the discharge densimetric Froude number given by

$$F_o = \frac{U_o}{\sqrt{g\,H_o\,\dfrac{\rho_a - \rho_o}{\rho_a}}} \qquad (4.31)$$

with the subscript a referring to ambient and o to discharge conditions. The form drag acting normal to the plume centerline is assumed similar to the drag on a solid body such that

$$F_D = \frac{1}{2}\sqrt{2}\,C_D HV|V|\sin^2\theta \qquad (4.32)$$

where C_D is a drag coefficient.

The interfacial shear forces are assumed to be similar to turbulent flow over a flat surface with a boundary layer thickness of $(2)^{1/2}H$ and a velocity equal to the

vector velocity difference between the plume and ambient current. Accordingly, the
X and Y components of this shear force reduce to

$$SF_X = C_F \left(\frac{1}{R_e H} \right)^{1/4} \int_0^{\sqrt{2}B} \Delta U_v^{3/4} \left[V \sin^2 \theta - U \cos \theta \exp\left(-\eta^2 / B^2\right) \right] d\eta \qquad (4.33)$$

$$SF_Y = -C_F \left(\frac{1}{R_e H} \right)^{1/4} \int_0^{\sqrt{2}B} \Delta U_v^{3/4} \left[V \cos \theta - U \exp\left(-\eta^2 / B^2\right) \right] d\eta \qquad (4.34)$$

where C_F is a friction coefficient and R_e is the jet discharge Reynolds number. The
value of C_F is determined by experiment.

The change in momentum flux includes the effects of the momentum of the
entrained ambient fluid, $V(dQ/ds)$, which acts in the X-direction. Equating the forces
to the change in momentum flux in the X and Y directions yields:

$$\frac{d}{ds}\left[(M + P)\cos\theta\right] = SF_X + F_D \sin\theta + V \, dQ/ds \qquad (4.35)$$

$$\frac{d}{ds}\left[(M + P)\sin\theta\right] = SF_Y - F_D \cos\theta \qquad (4.36)$$

Using Equations (4.17) and (4.31) for M and P, multiplying Equation (4.35) by
−sin θ, Equation (4.36) by cos θ, and combining yields an expression for the change
in flow direction,

$$\frac{d\theta}{ds} = \frac{SF_Y \cos\theta - SF_X \sin\theta - F_D V \sin\theta (dQ/ds)}{\dfrac{Q^2}{\pi BH} + \dfrac{\sqrt{\pi}}{2F_o^2} TH^2 B} \qquad (4.37)$$

Differentiating M and P, multiplying Equation (4.35) by cos θ and Equation (4.36)
by sin θ and combining yields the change in depth, Equation (4.38).

$$\frac{dH}{ds} = \left[SF_Y \sin\theta + SF_X \cos\theta + (V \cos\theta - 2Q/\pi BH)(dQ/ds) \right.$$

$$\left. -\left(\sqrt{\pi} BH^2 / 2F_o^2\right)(dT/ds) + \left(Q^2/\pi BH - \sqrt{\pi} H^2 / 2F_o^2\right)(dB/ds) \right] \qquad (4.38)$$

$$\left[\sqrt{\pi} \, THB / 2F_o^2 - Q^2 / \pi BH^2 \right]^{-1}$$

It is noted that this expression for change in depth is undefined when the denominator
is zero. Hence, results beyond this singularity are questionable.

Momentum in the lateral direction is included only indirectly through lateral spreading. It is assumed that the contributions to spreading by nonbuoyant horizontal jet mixing and buoyancy are independent of one another such that

$$\frac{dB}{ds} = \left(\frac{dB}{ds}\right)_{nb} + \left(\frac{dB}{ds}\right)_{b}$$

(4.39)

where the subscripts b and nb denote buoyant and nonbuoyant terms. The nonbuoyant spreading is found by writing Equation (4.38) without the buoyancy terms and assuming that

$$\left(\frac{dB/ds}{dH/ds}\right)_{nb} = (B/H)(dQ/ds)_{h}/(dQ/ds)_{v}$$

(4.40)

where $(dQ/ds)_h$ and $(dQ/ds)_v$ are the horizontal and vertical entrainment rates. This gives the nonbuoyant spreading rate:

$$\left(\frac{dB}{ds}\right)_{nb} = \frac{SF_Y \sin\theta + SF_X \cos\theta + \left(V\cos\theta - \dfrac{2Q}{\pi BH}\right)\dfrac{dQ}{ds}}{-\left(Q^2/\pi BH\right)\left[(dQ/ds)_v/(dQ/ds)_h + 1\right]}$$

(4.41)

The spreading due to buoyancy is assumed to be a function of the local excess density ratio, plume depth, and aspect ratio such that

$$\left(\frac{dB}{ds}\right)_b = \sqrt{\frac{2}{\dfrac{B}{H}F^2 - 1}}$$

(4.42)

It is noted that this also has a singularity. Due to B/H usually being large, this singularity is not encountered in most problems.

The preceding equations are sufficient to perform a stepwise integration along the plume. From the local conditions of the plume, dQ/ds is calculated. When this is known, dT/ds, $d\theta/ds$, and dB/ds are calculated. With these known, dH/ds can be calculated. These derivatives are integrated stepwise along the plume trajectory to give local values of X, Y, T, H, B, θ, and Q. Since PDS includes both jet entrainment and ambient vertical and horizontal turbulence, it gives smooth predictions through the near field to the far field. As entrainment fades, ambient turbulence takes over.

In order to start the integration within the developed zone where the above analysis is valid, starting conditions must be calculated. These are determined by a simplified analysis of the development zone assuming that the development length is given by the equation

$$S_i = 5.4\left(\frac{A^2}{F_o}\right)^{1/3}$$

(4.43)

where A is the discharge flow area and F_o is the discharge densimetric Froude number.

The main program, PDSG, prompts the user for input variables, initializes constants, nondimensionalizes the variables, and calls a standard scientific subroutine that performs the stepwise integration of differential equations by the Hamming predictor–corrector method. The predictions are saved on a file specified by the user and a temporary file used by the program to plot graphics. Plots are generated from the plume centerline trajectory, centerline concentrations (temperatures), widths, and assumed Gaussian distribution.

4.3.3 INPUT AND OUTPUT

Input is interactive with the terminal with the user answering prompts. After a run is completed, the user is given a chance to change any of the input variables and repeat a run. The input variables required are

1. A run title (one line).
2. The discharge flow rate (m³/s).
3. The discharge channel width (m).
4. The water depth in the discharge channel (m) — assumed rectangular. For pipes, approximate the discharge with an appropriate depth and width.
5. The ambient current (m/s).
6. Discharge angle relative to the current, degrees (0° is downstream and 90° is perpendicular to current).
7. Discharge water temperature (°C).
8. Ambient water temperature (°C).
9. Salinity (ppt) — both discharge and ambient assumed the same. If fresh water, use a small salinity such as 0.01.
10. Surface heat transfer rate. (You are given three choices: mild, average, or severe. This has little effect on temperatures or dilution in the near field.)
11. The distance downstream where simulations are to stop. (The program stops when this value is reached or you have 20 pages of output or when the excess temperature is .005 of the initial excess temperature.)
12. The name of a *nonexisting* file where the output is to be stored.

You are given the option of having the output (a) shown graphically on the screen, (b) sent in tabular form to a file, or (c) both. If the output is sent to a file, it consists of the following:

1. X, distance downstream in the direction of current (m).
2. Y, distance perpendicular to current (m).
3. Centerline excess temperature (temperature above ambient, °C).
4. Time of travel to this point along centerline (s).
5. Q/Q0, average plume dilution, total flow in plume divided by discharge flow.
6. QM/Q0, minimum dilution at plume centerline, which is one half the average due to the profiles assumed.
7. DEPTH, plume depth (m).

```
FLOATING WARM WATER JETS -- Feb 1998                                    PAGE    1
           PDSG sample run output


AMBIENT CONDITIONS     : TEMP. TA= 14.0 DEG. C ,;         VEL. =0.10 M/S
                         HEAT CONVECTION  =    1

DISCHARGE CONDITIONS : TEMP. = 19.0  C; DEPTH = 1.00 M. ; WIDTH = 20.00 M.
                       ANGLE  90.0 DEG  ;  DISCHARGE RATE =  20.00 CU-M/S
DISCHARGE DENSIMENTRIC FROUDE NO. =       10.95


   X(M.)       Y(M.)      EX.TEMP     TIME      Q/Q0    QM/Q0    DEPTH(M.)   WIDTH(M.)
                          (DEG. C)   (SEC.)    (DILU.)

    1.72       16.58       5.000    0.168E+02    2.00    1.00      1.39       36.77
    1.75       16.93       4.912    0.171E+02    2.04    1.02      1.44       36.91
    1.79       17.28       4.828    0.175E+02    2.07    1.04      1.48       37.04
    1.83       17.63       4.749    0.179E+02    2.11    1.05      1.52       37.18
                            .
                            .
                            .
  702.64      532.70       0.654    0.529E+04   15.27    7.63      4.62      379.81
  723.61      540.46       0.645    0.547E+04   15.49    7.75      4.63      385.31
  744.63      548.09       0.635    0.564E+04   15.72    7.86      4.65      390.75
  765.69      555.60       0.626    0.582E+04   15.94    7.97      4.66      396.14
  786.80      562.97       0.618    0.600E+04   16.17    8.08      4.68      401.47

   A R E A S  OF  E X C E S S   T E M P E R A T U R E    F O R
   PDSG sample run output


   EXC.  TEMP. (DEG. C)       AREA (SQ. M)

       0.25                  0.274E+06  PARTIAL AREA - AREAS VALID TO T =0.618 C
       0.50                  0.176E+06  PARTIAL AREA - AREAS VALID TO T =0.618 C
       0.75                  0.694E+05
       1.00                  0.245E+05
       1.25                  0.102E+05
                       .
                       .
       4.25                  0.327E+03
       4.50                  0.285E+03
       4.75                  0.242E+03
       5.00                  0.167E+03
```

FIGURE 4.17   Form of tabular output from PDSG.

8. WIDTH, plume width (m).
9. AREA, surface area within specified excess temperature isotherms (m²).
   These are given for each half degree of excess temperature. (If the simulation is terminated before a particular isotherm is reached on the plume centerline, only a partial area is given.)

Figure 4.17 is a partial tabular output for a sample case.

If graphical output is selected, a secondary screen similar to the one in UDKHG comes up giving you the opportunity to select the scale of the plot, contour lines to be plotted and their color, and the location of the legend. Color is given by number. Try a few to see what you get. Usually the default values are o.k. Follow the instructions on this screen to create the graphical representation of surface contours. For default settings, just pressing "d" will display the graph. Figure 4.18 is a sample graphical plot showing surface temperature (or concentration) contours for discharge normal to a moderate current. The graph ends where calculations ended. Contour

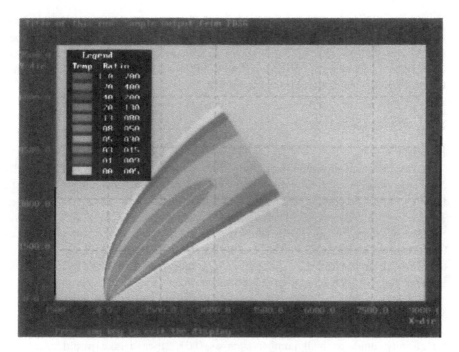

FIGURE 4.18   Sample graphical output from PDSG showning contour bands.

lines are labeled as excess temperature above the ambient divided by the discharge excess temperature. These can also be considered as concentration ratio (C/C_0), if the surface heat transfer coefficient is low or the discharge excess temperature is small. (Don't use zero or the program will blow up. A one degree difference is o.k. and will have little effect on the plume.)

The graphical screen created by PDSG can be converted into a bitmap file and printed out anywhere you desire by copying it to the PAINT or other graphic application in Windows 95 or NT. To do this, create the desired graph using PDSG, while viewing the graph press <control><print scrn> (both together), then go to PAINT and select FILE PASTE. You can resize the graph and save it to the file you desire.

After the run, you are given the option to (a) display the screen again with different scale or contours, (b) rerun the case and change some values, or (c) exit. If you select to rerun the case, you are given a screen with input values where you can change any or all and rerun.

The program has been tuned to a wide variety of data and agrees with the average of these data as outlined in the reference given at the beginning of this paper. It may not agree exactly with each specific case but should give reasonable answers if boundaries are not encountered. *The program does not keep track of receiving water boundaries*, i.e., shore or bottom. The user must check to see that the plume does not attach itself to either shore or bottom by following the trajectory, plume width, and depth. Simulations beyond these attachment points are in error due to changes in entrainment.

The image method can also be used with PDSG to simulate discharge parallel to the bank or discharge into a strong current that causes the plume to hug the near shore. This can be done by doubling the discharge rate and width. In the output, consider only one half the predicted plume and assume that the centerline (highest temperature or concentration) is at the shore.

4.4   PDSG EXAMPLES

Example 4.6

Condenser cooling water is discharged into a large lake at a rate of 20 m^3/s. The discharge channel is 20 m wide and the water depth in the channel as it enters the lake is 1.0 m. The cooling water and lake temperatures at the time of the study are 33°C and 10°C, respectively. There is a residual current along the shore of 0.01 m/s. Surface heat transfer can be considered as moderate. Regulations do not allow excess temperatures above 2.0°C outside a 3000-m radius from the discharge point or a total surface area within the 2.0°C contour greater than 1,000,000 m^2. Use PDSG to determine if this discharge exceeds regulations

Solution:

Run PDSG using the values given in the problem statement. When prompted for salinity, enter a small number such as 0.01, since a lake would usually be freshwater. When prompted for a region of interest, enter a number greater than 3000 since you want answers at least that far. Larger numbers just give more output and make the graphed region bigger. Give an output file name that doesn't exist and select BOTH for both graphic and file output. Figures 4.19 and 4.20 show the tabular and graphical output for this example. As can be seen, the 2° excess isotherm extends to about 5500 m from the discharge with an enclosed surface area of 4,680,000 m^2. Both of these values exceed regulations. Mixing could be improved, resulting in a shorter distance to the 2° isotherm by discharging through a smaller channel at higher exit velocity if that is possible.

Example 4.7

Treated industrial waste is discharged at right angles to a river from a 3 ft diameter pipe near the surface of the river. The river is 300 m wide and 10 m deep and has an average current of 0.2 m/s. The effluent discharge rate is 90 ft^3/s (2.55 m^3/s) and is essentially at the same temperature as the river. Regulations are such that a minimum centerline dilution of 20 is required before the plume reaches 100 m downstream.

Solution:

A 3 ft diameter pipe can be represented by a square duct with 0.81 m sides. Run PDSG and enter 0.81 for both width and depth. Enter the other values when prompted. For discharge and ambient temperatures, enter 20 and 19, respectively. This will give a plume with little or no buoyancy and the column in the output labeled EX. TEMP will represent the centerline concentration ratio, C/C_o. For salinity

```
FLOATING WARM WATER JETS -- Feb 1998                              PAGE   1
         EXAMPLE 4.6


AMBIENT CONDITIONS    : TEMP. TA= 10.0 DEG. C ,;          VEL. =0.10 M/S
                        HEAT CONVECTION  =    2

DISCHARGE CONDITIONS : TEMP. = 33.0  C; DEPTH = 1.00 M. ; WIDTH = 20.00 M.
                        ANGLE  90.0 DEG  ;  DISCHARGE RATE = 20.00 CU-M/S
DISCHARGE DENSIMENTRIC FROUDE NO. =    4.50
```

X(M.)	Y(M.)	EX.TEMP (DEG. C)	TIME (SEC.)	Q/Q0 (DILU.)	QM/Q0	DEPTH(M.)	WIDTH(M.)
2.27	22.25	23.000	0.229E+02	2.00	1.00	1.11	47.43
2.31	22.59	22.663	0.232E+02	2.03	1.01	1.14	47.59
2.35	22.94	22.349	0.236E+02	2.06	1.03	1.18	47.76
2.38	23.29	22.057	0.240E+02	2.09	1.04	1.21	47.93
2823.15	1994.17	2.994	0.207E+05	12.49	6.25	1.51	921.50
2844.26	2001.55	2.978	0.208E+05	12.54	6.27	1.52	924.57
2865.38	2008.90	2.962	0.210E+05	12.59	6.30	1.52	927.63
2886.51	2016.20	2.946	0.212E+05	12.64	6.32	1.52	930.68
2907.66	2023.48	2.930	0.213E+05	12.69	6.35	1.52	933.72
2928.82	2030.71	2.914	0.215E+05	12.74	6.37	1.53	936.75
2949.99	2037.92	2.898	0.217E+05	12.79	6.40	1.53	939.77
2971.17	2045.08	2.883	0.219E+05	12.84	6.42	1.53	942.78
2992.36	2052.22	2.867	0.220E+05	12.89	6.45	1.53	945.79
3013.56	2059.32	2.852	0.222E+05	12.95	6.47	1.54	948.78
3034.78	2066.38	2.837	0.224E+05	13.00	6.50	1.54	951.77
4107.47	2380.79	2.179	0.313E+05	15.76	7.88	1.69	1091.77
4129.12	2386.38	2.168	0.315E+05	15.82	7.91	1.69	1094.42
4150.78	2391.94	2.157	0.316E+05	15.88	7.94	1.70	1097.05
4172.45	2397.48	2.145	0.318E+05	15.94	7.97	1.70	1099.68
4194.12	2403.00	2.134	0.320E+05	16.01	8.00	1.70	1102.31

```
     A R E A S  OF  E X C E S S   T E M P E R A T U R E   F O R
         EXAMPLE 4.6


     EXC.  TEMP. (DEG. C)      AREA (SQ. M)

         1.15               0.371E+07  PARTIAL AREA - AREAS VALID TO T =2.134 C
         2.30               0.203E+07
         3.45               0.858E+06
```

FIGURE 4.19 Sections of tabular ouput from PDSG for Example 4.6.

enter a small number and select low surface heat transfer. For the region of interest, enter something greater than 100, say 500. For output, select BOTH. The graphical output will be particularly important to see if the plume attaches to either bank. Figures 4.21 and 4.22 are modified versions of the tabular and graphical output. From the tabular output, it is found that the centerline dilution is about 25 when the plume reaches 100 m downstream, which is greater than the required value. The plume depth at this point is about 8 m, which is less than the river depth, meaning the plume has not become attached to the bottom. The graphical output shows that the plume does not hit either the near or far shore of the 300 m wide river.

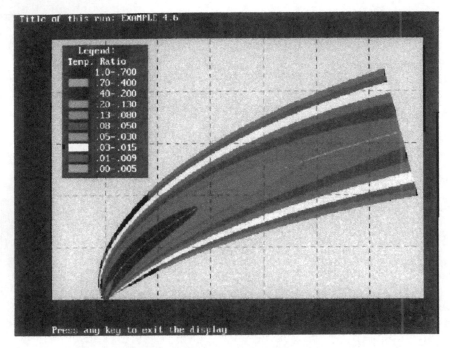

FIGURE 4.20   Graphical ouput from PDSG for Example 4.6.

Example 4.8

What would the dilution be at 100 m downstream for Example 4.7 if the velocity were 1.0 m/s and high enough to cause the plume to attach to the near shore?

Solution:

Reenter the values given in Example 4.7, but this time use the image method. To do this, use a width of $2 \times 0.81 = 1.62$ m and a discharge rate of $2 \times 2.55 = 5.1$ m³/s, and to be conservative, enter a discharge angle of zero. This makes it attach to the shore immediately, ignores initial river interaction, and gives the most conservative answer. Select output to a file. From the output, you will see that the centerline dilution 100 m downstream is only about 11. This is a result of less interaction with the current and reduction of entrainment from the near shore. If you also selected graphical output, remember to only consider one half of the plume shown. The other half is the image half causing symmetry and the maximum concentration to be along the shore.

Problems

4.1 Repeat Problem 3.1 using UDKHG. (*Hint:* If the negative density gives you a problem, turn your coordinate system upside down and reverse the effluent and ambient densities.)

4.2 A "Y" shaped diffuser discharges 4.38 m³/s through 40–20 cm ports spaced 4 m apart. Twenty of the ports are on one branch of the Y and 20

```
FLOATING WARM WATER JETS -- Feb 1998                              PAGE   1
          EXAMPLE 4.7


AMBIENT CONDITIONS     : TEMP. TA= 19.0 DEG. C  ,;        VEL. =0.20 M/S
                         HEAT CONVECTION  =    1

DISCHARGE CONDITIONS : TEMP. = 20.0  C; DEPTH = 0.81 M. ; WIDTH =  0.81 M.
                       ANGLE  90.0 DEG  ;  DISCHARGE RATE =   2.55 CU-M/S
DISCHARGE DENSIMENTRIC FROUDE NO. =      96.74


     X(M.)      Y(M.)     EX.TEMP     TIME      Q/Q0    QM/Q0   DEPTH(M.)  WIDTH(M.)
                          (DEG. C)    (SEC.)    (DILU.)


      0.05       0.97     1.000   0.249E+00      2.00    1.00      1.26       1.33
      0.05       1.03     0.976   0.266E+00      2.05    1.02      1.26       1.39


     62.55      53.11     0.053   0.206E+03     37.80   18.90      6.65      39.54
     66.39      54.41     0.051   0.221E+03     39.18   19.59      6.75      40.68
     70.24      55.65     0.050   0.236E+03     40.55   20.27      6.84      41.79
     74.11      56.85     0.048   0.251E+03     41.91   20.95      6.93      42.88
     77.99      58.00     0.047   0.267E+03     43.25   21.63      7.01      43.95
     81.89      59.12     0.045   0.282E+03     44.59   22.29      7.10      45.00
     85.79      60.20     0.044   0.298E+03     45.92   22.96      7.18      46.03
     89.70      61.25     0.043   0.313E+03     47.23   23.62      7.25      47.04
     93.62      62.26     0.041   0.329E+03     48.54   24.27      7.33      48.03
     97.55      63.24     0.040   0.345E+03     49.85   24.92      7.40      49.01
    101.49      64.20     0.039   0.361E+03     51.14   25.57      7.47      49.97
    105.43      65.12     0.038   0.378E+03     52.43   26.22      7.54      50.91
    109.38      66.03     0.037   0.394E+03     53.72   26.86      7.61      51.84
    113.33      66.90     0.037   0.410E+03     54.99   27.50      7.68      52.76
    117.29      67.76     0.036   0.427E+03     56.27   28.13      7.74      53.67
    121.25      68.59     0.035   0.444E+03     57.53   28.77      7.80      54.56
    125.22      69.40     0.034   0.460E+03     58.80   29.40      7.87      55.44
```

FIGURE 4.21   Sections of tabular ouput from PDSG for Example 4.7.

are on the other. The effluent has a density of 1000 kg/m³ and is discharged 20° up from the horizontal. The ports discharge perpendicular to the diffuser pipe. Because of the Y shape, the prevailing current makes an angle of 45° with the diffuser. Determine the dilution at the end of the zone of initial dilution (when the plume surfaces or is trapped) if the ports are 15 m deep and the ambient conditions are the same as given in Problem 3.3c.

4.3 Warm water is discharged into a large river from a perpendicular side channel. The channel is 3 m wide and 0.5 m deep. The discharge rate is 2.6 m³/s at 41°C. The ambient is at 20°C and has an average current of 0.15 m/s. Determine the location in the river where the temperature is 2° warmer than the river if average surface heat transfer is used and the surface area within the 2° isotherm.

4.4 30 mgd of treated industrial waste are discharged into a shallow river through a diffuser that has eight 8-in. ports. Discharge is horizontal and 90° from the current (cross current). If the river is 2.44 m deep and has an average velocity of 0.18 m/s, use the image method to analyze one of the ports and determine the dilution 10 and 100 m downstream. Ignore density differences.

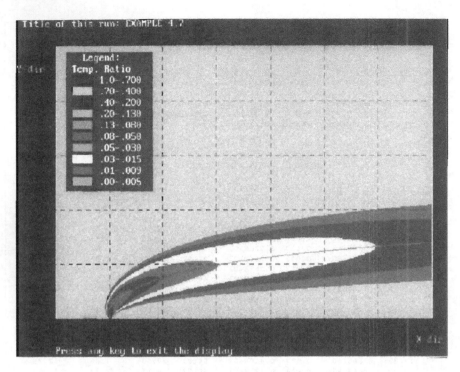

FIGURE 4.22   Graphical ouput from PDSG for Example 4.7.

5 Empirical and Length Scale Models

5.1 BASIC CONCEPTS

The basic logic behind length scale models is that the variables involved with the forces taking part in the mixing process can be arranged in groups that have the dimensions of length. The magnitude of these length scales indicate how important a particular process is. Ratios of these length scales form dimensionless groups that can be used in scaling the results of laboratory studies to full-sized field processes such as Reynolds number is used in frictional flow. Experiments are performed in the laboratory over a range of variables that cover the expected range in the field. The results of these experiments are used to determine empirical relations between the dimensionless groups. These relationships can then be used to predict the behavior of similar configurations in the field.

One single relation is usually not enough to accurately describe experimental behavior over the complete range of interest. For example, let's assume that the variables involved in a particular discharge can be grouped into three length scale variables, L1, L2, and L3, and dilution, Q/Qo. It is found that using the ratios of L1/L2 and L3/L2 cause the dilution to collapse together, as shown on Figure 5.1. Equations are found using L1/L2 and L3/L2 as independent variables and dilution as the dependent variable that fall through the center of the data. In the example of Figure 5.1, however, it takes two equations to fit the data for both high and low values of the ratio of L1/L2. For example, region 1 could represent the near field and region 2 the far field where mixing mechanisms are completely different. When predictions are made using these equations, the value of L1/L2 is calculated from the user's data. This determines which equation to use. The appropriate equation along with the value of L3/L2 is used to determine the dilution.

This method usually works well as long as the configuration is the same for the user and the one used to generate the equations, and the user's data falls within the range where the experiments were made. There are, however, several cautions. First, using the equations to predict behavior beyond the range where the original data were taken (extrapolation) can lead to large errors. Second, the equations used to best fit the data may not be continuous, as shown in Figure 5.1, and predictions on one side of a boundary will be different from predictions on the other side of the boundary. And lastly, it is usually impossible to have all conditions in the laboratory be the same as those in the field, such as boundary effects and turbulence. Users of this method should be aware of the limitations and use the results with caution in questionable cases. The advantage is that the equations are easy to evaluate and

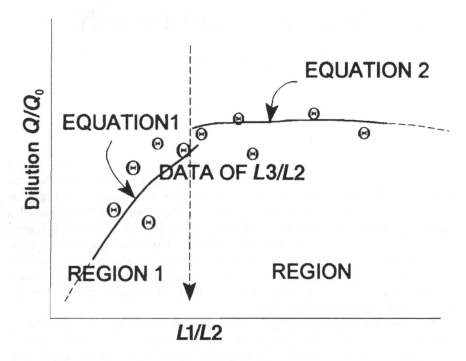

FIGURE 5.1 Sample graph showing how length scale ratios are used to determine empirical equations.

often include physical effects such as boundary interactions that are not easily included in mathematical models such as UDKHG and UM.

The main empirical models discussed in this book are the ocean diffuser model, RSB,[24] and the expert system of models called CORMIX.[25]

5.2   RSB MODEL

The RSB model is an empirical length scale model designed for ocean outfalls based on the experimental studies by Roberts, Snyder, and Baumgartner.[26-28] These studies were conducted on a diffuser with T-shaped risers spaced at equal distances along the diffuser, as shown in Figure 5.2. The risers had two ports discharging opposite each other in a horizontal direction. The ambient was linearly stratified and the current was uniform. Various angles of the diffuser line relative to the ambient current were studied.

In the RSB model, volume, momentum, and buoyancy fluxes are defined as

$$q = Q/L; \; m = U_j q; \; b = g'q$$

respectively, where Q is the total discharge rate in the diffuser, L is the diffuser length, U_j is the nozzle jet velocity, and g' is the reduced buoyancy given as gravity

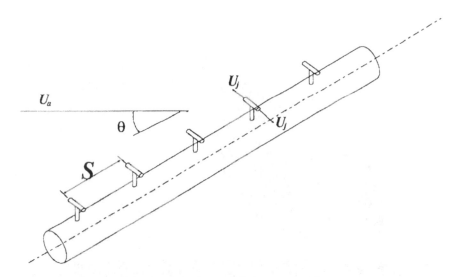

FIGURE 5.2   T-shaped diffuser used in experiments to create data for RSB model.

times the density difference between ambient and effluent divided by the ambient density:

$$g' = \frac{g(\rho_a - \rho_o)}{\rho_a} \tag{5.1}$$

The Brunt–Vaisalla frequency, N,

$$N = \pm \sqrt{\frac{g}{\rho_a} \frac{|d\rho|}{dz}} \tag{5.2}$$

is defined as the square root of gravity times the density gradient in the ambient over ambient density. It is sometimes called the stratification parameter, since it relates to the stratification gradient in the ambient. These fluxes and N are then combined to form the following length scales:

$$l_q = q^2/m; \qquad l_b = b^{1/3}/N; \qquad l_m = m/b^{2/3}$$

Other lengths of interest are the port spacing, s, the distance to the end of initial dilution, X_i, the height to the top of the wastefield layer, Z_e, the thickness of the waste field after trapping, h_e, and height above discharge to the point of minimum dilution, z_m, as shown in Figure 5.3 The final variable of interest is the average dilution, $S = Q/Q_0$, the ratio of flow in the plume at the end of initial dilution divided by the discharge flow. The minimum dilution at the point of maximum concentration in the plume was determined Robert's experiments to be about $S_m = S/1.15$. (Note that this is different than the values used in other methods.)

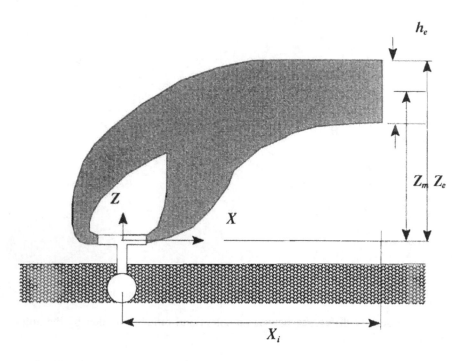

FIGURE 5.3   Sketch of plume showing definition of various lengths used in RSB.

The primary dimensionless groups formed as independent variables are

$$l_m/l_b, \qquad s/l_b, \qquad F = U_a^3/b, \text{ and diffuser angle}$$

where U_a is the ambient velocity. The functional relationships between these variables and desired dependent variables were found as mathematical expressions in some cases and as curves fit to graphical results in others. In general, four dimensionless dependent ratios can be expressed as functions of four independent values as follows:

$$\frac{S_m q N}{b^{2/3}}, \frac{z_e}{l_b}, \frac{h_e}{l_b}, \frac{z_m}{l_b} = f\left(\frac{l_m}{l_b}, \frac{s}{l_b}, F, \Theta\right)$$

where $F = U_a^3/b$ and is known as Robert's F.

The empirical expressions that represent the functional relationships are programmed in RSB. They are valid for $s/l_b < 1.92$, $l_m/l_b < 0.5$, $F < 100$, and $0 \le \theta \le 90$, which are typical of ocean discharges. The results are good within the range of variables used to determine the expressions. Outside this range, the program gives a warning that the answers are questionable. It selects the ones to use depending on the ratio of the independent functions. In some cases, certain independent length scale ratios are not important. For example, the minimum dilution and distance to the top of the waste field for buoyant discharge from a diffuser with closely spaced ports that is perpendicular to a current into a stratified ambient and has low momentum flux is given by

$$\frac{S_m qN}{b^{2/3}} = 2.19F^{1/6} - 0.52 \tag{5.3}$$

$$\frac{z_e}{l_b} = 2.5F^{-1/6} \tag{5.4}$$

for $0.1 < F < 100$.

The length of the initial dilution zone, X_i, was defined as the distance where small-scale jet turbulence changed to larger scale ambient turbulence. This was done visually using shadow graphs and assuming that jet mixing was highest where turbulence scales were small. This distance is somewhat arbitrary, since there is no exact line where this change takes place. It also varies from case to case. So if a user is interested in the dilution at a fixed mixing zone boundary, this model cannot be used, since it only gives answers at the distance X_i from the discharge it determines. Intermediate values and trajectory are not given. A careful review of the length scales will show that l_b is zero for nonstratified ambients. This requires that different relations be used for this case, since the plume will not be trapped but will collapse at the surface. These are also included in the model.

The RSB program is easy to run using the EPA's PLUMES interface discussed in Chapter 3. RSB is included within the PLUMES package and can be downloaded from the World Wide Web as given in Chapter 3. RSB can also be run directly from the Windows version of RSB, RSBWIN.* Figure 5.4 is a sample of the input screen of RSBWIN. Figure 5.5 is a sample of the output screen from RSBWIN. It is noted that RSB gives the values of the length scale ratios determined from the input and the results of its predictions. The predictions include distance from discharge to the top of the waste field, the submergence distance below the surface to the top of the waste field, the thickness of the waste field, the height to the point of minimum dilution (maximum concentration), the length of the initial mixing region, and the average and minimum dilutions. It does not give intermediate values.

Because RSB is limited to $l_m/l_b < 0.5$ (high buoyancy relative to discharge momentum), it is not suited for discharges into inland waters and streams. Its application is mainly for freshwater discharge into the ocean where large density differences between discharge and ambient occur and where the zone of initial dilution defined by the program can be used as the mixing zone. Any variation in ambient stratification can be entered into the RSB program, but it reduces the data to an equivalent linear stratification by using the ambient density at discharge level and the density at the maximum height of rise in evaluating the density gradient in the Brunt–Vaisalla frequency, N. Since the height of rise is not known to start with, it must be determined by iteration. For most stratification variations found in nature,

---

* The Windows version can be downloaded from www.water.ce.gatech.edu/Faculty/P.Roberts/projsoft.html.

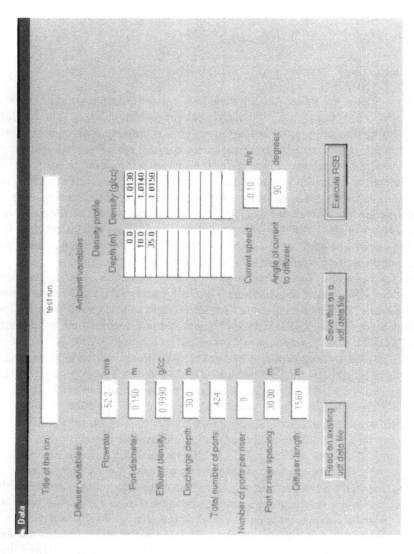

FIGURE 5.4 Sample input screen for RSBWIN. Windows version of RSB.

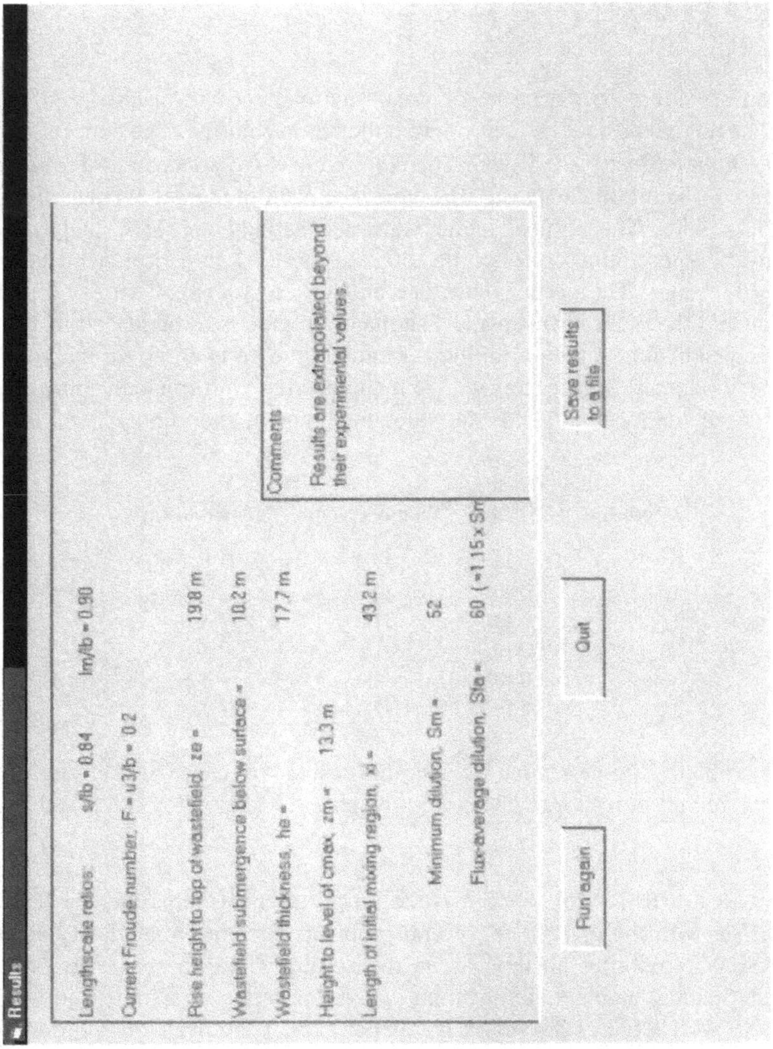

FIGURE 5.5 Sample output screen for RSBWIN.

this leads to a conservative prediction and gives initial dilution lengths and dilutions lower then actual. If you want to use RSB on a diffuser with ports on only one side, as is simulated with UM and UDKHG, the spacing between risers simulated in RSB should be double the spacing between the actual ports. This is because RSB assumes two ports for each riser.

5.3   RSB EXAMPLES

Example 5.1

A diffuser is to be placed in a coastal environment and discharge 40 mgd of treated municipal waste. The controlling pollutant is chlorine. The concentration of chlorine in the effluent is 0.5 o/oo. The required average concentration of chlorine at the end of the mixing zone is 0.011 o/oo. The resulting target average dilution is 0.5/0.011 = 45.5. The effluent temperature and salinity are 15°C and 0.1 o/oo, respectively. A preliminary design has 30 6-in. ports, 2 to a riser, with the risers placed 20 ft apart. The ports appose one another on the risers with one pointing upstream and the other downstream. The manifold pipe is to be placed in 100 ft of water, perpendicular to the prevailing current and the ports are to discharge up 20° from the horizontal. The nozzles will be manufactured with rounded approaches. A survey of the site location yields the following ambient conditions:

Depth (m)	Velocity (ft/s)	Salinity (o/oo)	Temperature (°C)
0	0.20	30.0	10.0
10	0.20	30.5	9.5
15	0.20	31.0	9.5
20	0.15	31.5	9.0
30	0.10	32.0	8.5
35	0.10	32.0	8.5

The problem is to determine whether this design will cause the chlorine content within the plume to reach 0.011 before it reaches the surface or is trapped.

Solution:

We will use RSBWIN, since it is the most current. Comparing the RSBWIN input screen with the given data, several problems become immediately apparent. First, RSBWIN requires all input be in metric units, and no conversion utility is provided. Second, ambient stratification is given in terms of salinity and temperature.* RSBWIN requires density in grams per cubic centimeter. In addition, RSBWIN will only accept one ambient current value, so a representative average must be selected. And lastly, RSBWIN has no provisions for variations in vertical discharge angle. The metric units and density problems can be eased by using the

---

* A program called DENSITY that approximates water density from temperature and salinity can be downloaded from the World Wide Web from www.engr.orst.edu/~davisl.

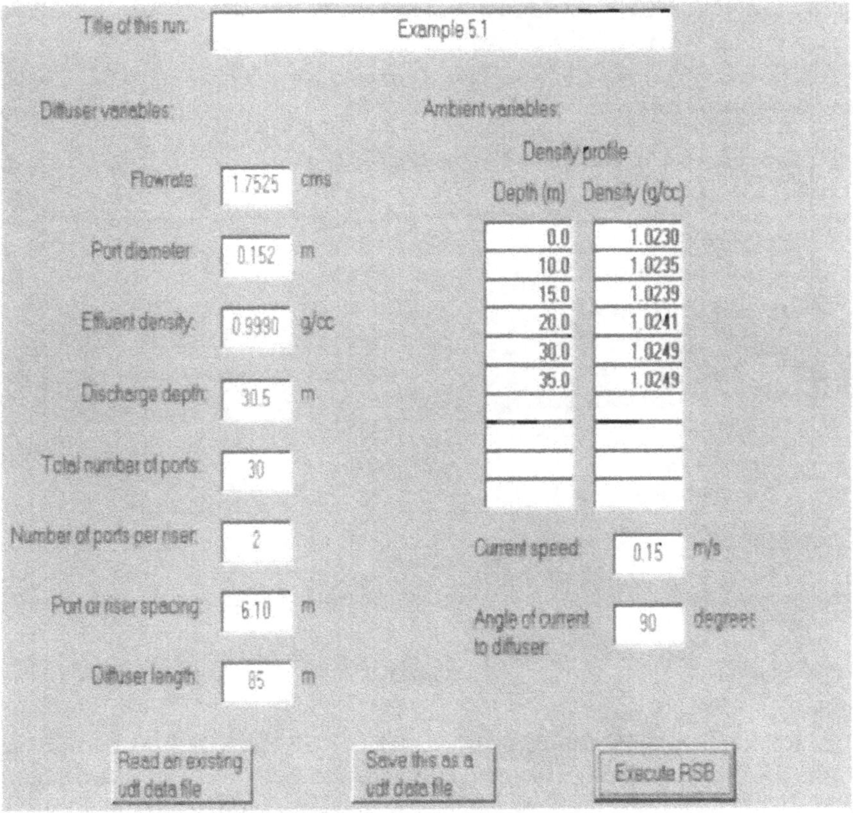

FIGURE 5.6   RSBWIN input screen for Example 5.1.

PLUMES interface to perform the conversions, if you don't have something you prefer. The values can then be entered into RSBWIN. That RSB will not reflect the 20° upward discharge angle is not a serious problem, since it doesn't give details near the discharge anyway and slight variations in vertical discharge angle will not affect values at the end of the zone of initial dilution very much. Figure 5.6 shows how the RSBWIN input screen would look for this problem with the ambient current set at 0.15 m/s. Figure 5.7 is the resulting output. As the output shows, RSB predicts that the plume will be trapped with the upper edge of the waste field 12.0 m below the surface. The zone of initial dilution as defined by RSB extends 50.4 m downstream where the average dilution is 100. Since the required dilution was 45.5, this configuration would be satisfactory.

Example 5.2

The previous example showed that a current of 0.15 m/s perpendicular to the diffuser would produce the desired dilution. In this example, let's see if the same diffuser will work if there is a small current of 0.01 m/s parallel to the diffuser.

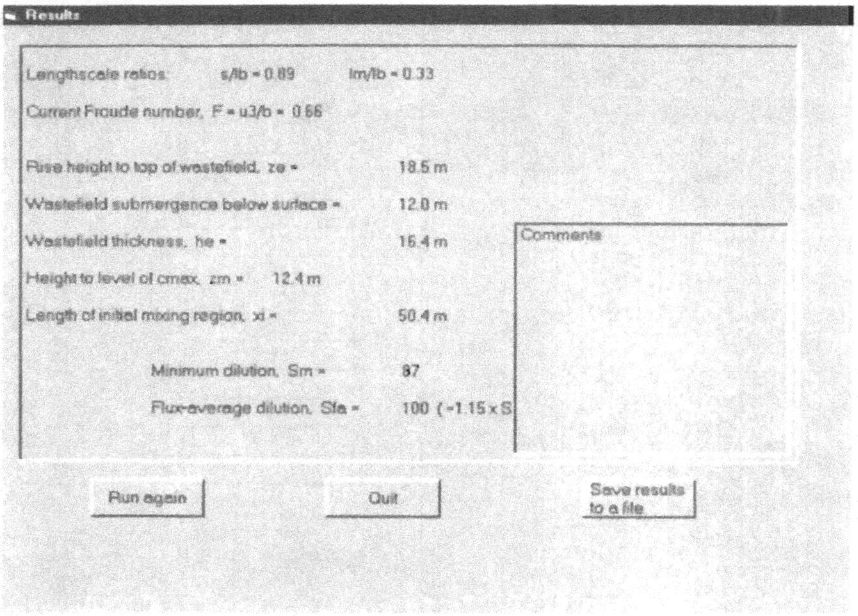

FIGURE 5.7   RSBWIN output screen for Example 5.1.

Solution:

From the RSBWIN Results window, select RUN AGAIN. This puts you back to the input screen. Change the current magnitude and direction and execute RSB again. Figure 5.8 is the resulting output screen. In this case, the length of the zone of initial dilution is only 13.6 m where the average dilution is 64. The plume is still trapped. These values indicate that the proposed diffuser would give satisfactory dilutions for a range of currents. One of RSB's strong points is the ability to predict diffusers parallel to the ambient current.

5.4   CORMIX MODELS

CORMIX[25] is a software system that incorporates an expert system interface with a number of hydraulic models. The program along with a user's manual can be downloaded from the Internet.* CORMIX is designed to handle a wide variety of discharge configurations and ambients including single and multiple port diffusers, submerged and surface discharge, positive and negative buoyant discharges, diffusers with unidirectional, fanned, alternating ports or risers with multiple ports, open oceans, lakes, rivers, and estuaries. It can also consider tidal varying ambients in a quasi-transient approach. CORMIX is inherently steady-state, but it will consider the rate of change of ambient conditions and account for re-entrainment due to tidal reversal.

---

* CORMIX and the user's manual can be downloaded from ftp://ftp.epa.gov/epa_ceam/wwwhtml/cormix.htm or from Dr. Robert Doneker, who manages CORMIX, at http://www.ese.ogi.edu/ese_docs/doneker.html.

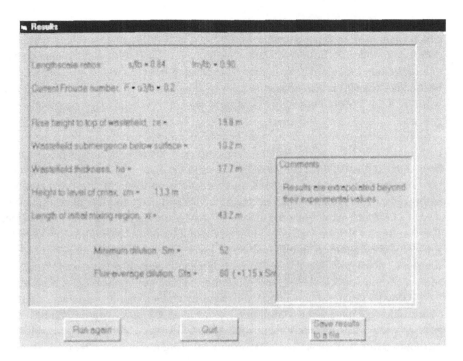

FIGURE 5.8 RSBWIN output screen for Example 5.2.

The expert system interface is highly user-interactive: you are actually interviewed by the program about the proposed project. You answer a series of questions regarding the diffuser, the ambient, receiving water geometry, and regulations. It gives you the opportunity to receive explanations at several steps during the process regarding the particular input required. These explanations can be bypassed once you understand what is wanted by selecting NO to all the prompts or by selecting the fast input version from the opening screen. Because of the extensive explanations that can be obtained from CORMIX, it can be used as a training tool for people not familiar with environmental discharges and regulations.

Once all the questions have been answered and the data entered, CORMIX determines a number of length scales. These are used to determine the FLOW CLASS. FLOW CLASS is the system used by CORMIX to specify which type of plume is expected to develop as a result of the discharge configuration and ambient. For example, is the flow jet-like or plume-like, is stratification sufficient to dominate, is there a weak current or strong current, is the receiving water deep or shallow, is the alignment parallel or perpendicular, is discharge positively buoyant or negatively buoyant, is there Coanda bottom attachment or wake attachment, is there an upstream wedge, etc. There are something like 66 flow classes for submerged discharge alone. CORMIX then runs the hydraulic models it has determined are appropriate for this flow class in sequence and patches them together.

The philosophy behind CORMIX is good, but the wide range in discharge and ambient conditions considered cannot be categorized by even 66 flow classes without having some discontinuities in predictions, as discussed using Figure 5.1. A very small change in an input variable may result in a different flow class with quite different predictions. Both flow classes may be realistic, but the sudden change from one to the other may not be. Another flow class in between the two would result in a smoother transition. An infinite number of flow classes is unrealistic, so you must work with the discontinuities that occur and realize that the actual transition is more gradual.

These hydraulic models consist of near field integral models similar to the PDSG, UM, and UDKHG models, length scale empirical models similar to RSB, and far field ambient passive diffusion models similar to the ones in PLUMES. Once hydraulic calculations are made, CORMIX creates reports on its predictions in a summary report regarding regulations and plume behavior and in a complete hydraulic output report where the results of each phase of the hydraulic calculations are reported. The results can also be viewed graphically using an optional graphics package that shows the predictions in simple line graph form. The fact that an inexperienced user can run CORMIX and get answers does not mean that anyone can do a mixing zone analysis. Some knowledge of fluid mixing and some common sense is a must.

CORMIX requires large amounts of memory and will not run on every computer. Many older computers must be booted from a floppy disk without Windows and have no programs in resident memory before CORMIX can be installed or loaded. A bootable floppy disk can be created by formatting it with the FORMAT/S option.

5.4.1   CORMIX INPUT

If you are serious about becoming a CORMIX user, you should download the complete CORMIX manual and review it before attempting to make runs for a project; however, this book provides what is needed to get started. The opening screen for CORMIX allows the user to

- Start a new session
- Rerun and modify a former CORMIX run
- Redisplay the summary results of a former run
- Post processor for graphics and far field calculations
- Manage CORMIX files
- Change from regular to fast CORMIX input
- Select whether flow class is displayed or not
- Quit CORMIX

You are not able to use a mouse pointer in the present version of CORMIX. All movement is done from the keyboard using arrow buttons. A Windows version of CORMIX is under development that should make input a lot easier. For a new session, CORMIX requires four sets of input data. After entering each set of data, you are given the option to accept or re-enter the data. If you make a mistake entering data and have gone on to the next prompt before discovering it, you have to complete entering data for that set and then reject the input at the end. You can then re-enter all the data for that section again with the values you want. Table 5.1 is a worksheet

TABLE 5.1
Worksheet for Entering CORMIX Input Data

CORMIX INPUT

Site Name _____

Design Case _____

DOS File Name _____ (w/o extension)

AMBIENT DATA

Water body depth	_____ m	If bounded: Width		_____ m
Depth at discharge	_____ m	Appearance	1/2/3	
If Steady: Ambient flowrate	_____ m³/s	Ambient velocity		_____ m/s
If Tital: Tidal Period	_____ hr	Max. Tidal velocity		_____ m/s
At time ____ hr before/after/at slack:		Tidal velocity at this time		_____ m/s
Manning's n	_____ or:	Darcy's f		_____
Wind speed	_____ m/s			
Density data		Units: Density...kg/m³, Temperature .. °C		
Water body is	fresh/salt water			
If uniform		Temperature or density (salt water)		_____
If Stratified		Temperature or density at surface		_____
Stratification type	A/B/C	Temperature or density at bottom		_____
If B or C Pycnocline height	_____ m	If C: Temp or Tensity at jump		_____

DISCHARGE DATA: Specify whether CORMIX1 or 2 or 3

```
CORMIX1 -- Submerged Single Port
Nearest bank is on    left/right          Distance to nearest bank        _____ m
Vertical angle THETA  _____ °         Horizontal angle SIGMA          _____ °
Port diameter         _____ m or      Port area                       _____ m²
Port height           _____ m
```

```
CORMIX2 -- Submerged Multiple Port Diffuser
Nearest bank is on    left/right          Distance to one endpoint        _____ m
Diffuser length       _____ m               to other endpoint        _____ m
Total number of ports _____ m         Port height                     _____ m
Port diameter         _____ m         Contraction ratio               _____
Diffuser arrangement or type  unidirectional/ staged / alternating / vertical
Alignment angle GAMMA _____ °         Horizontal angle SIGMA          _____ °
Vertical angle THETA  _____ °         Orientation angle BETA          _____ °
```

```
CORMIX3 -- Surface Discharge
Discharge located on  left/right  bank    Configuration  Flush/Protruding/Co-flowing
Horizontal angle SIGMA _____ °        If protruding: Distance from bank  _____ m
Depth AT discharge    _____ m         Bottom slope                    _____
If rectangular: Width _____ m         If circular:   Diameter          _____ m
           Depth      _____ m                        Pipe invert depth _____ m
```

Effluent: Flow rate	_____ m³/s or	Effluent velocity	_____ m/s
Density or temperature	_____ kg/m³ or °C		
Heated discharge?	Yes/no	If yes: Surface heat loss coefficient	_____ W/m²·C
Concentration units	_____	Effluent concentration	_____
Conservative substance	Yes/no	If no: Decay coefficient	_____ /day

MIXING ZONE DATA:

Is effluent toxic	Yes/no	If yes: CMC_____		CCC_____	
WQ standard?	Yes/no	If yes: value of standard			
Mixing zone Specified?	Yes/no	If yes: distance _____ or	width	_____	% or m
		or area	_____	% or m²	
Region of interest	_____ m	Grid intervals for display		_____	

with all the required input variables and options. It helps to copy this page and fill in the blanks before entering the data for each case. CORMIX checks the consistency of your data. If values you enter are inconsistent, it flags you and will not go on until you correct the inconsistency.

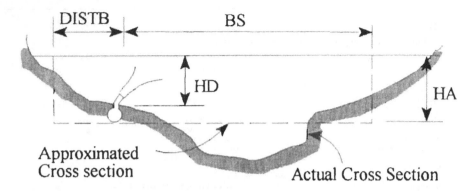

FIGURE 5.9   Cross section of bounded ambient with CORMIX rectangular simplification.

5.4.1.1   Project Data

The first set of data is just project identification such as **site name, design case,** and a **file name** where the input and output are to be stored. This name should not have an extension, since CORMIX automatically adds different extensions depending on the file. For example, in the file SIM\fn.CXn is the output prediction report for the run where SIM is a subdirectory in the CORMIX main directory, fn is the file name you gave, and CXn is an extension with "n" being 1, 2, or 3 referring to which CORMIX model you ran, i.e., CORMIX1, CORMIX2, or CORMIX3. In the file SIM\CXn\fn.CXC is a list of all your responses to the prompts for the run and related calculations CORMIX makes during the input process. In this case, CXn is one of three subdirectories in the subdirectory SIM with n again being 1, 2, or 3, and CXC is an automatic extension. It is not a good idea trying to edit this file to change a case to rerun. You might miss related values that CORMIX calculates during the input process. It is better to use the "rerun and modify case" option in the opening menu and re-enter the values in the section of input you want changed.

5.4.1.2   Ambient Data

The second set of input data is for ambient conditions where you enter the type of receiving body, stratification, and currents. You must first enter whether the receiving body is **bounded** or **unbounded**. Bounded is for cases where the plume may attach to both shores such as a river or narrow estuary. Unbounded is for cases where the plume will attach to only one shore, if any, such is in coastal waters or large lakes. For the bounded case, CORMIX approximates the receiving water with a rectangular cross section with an **average depth,** HA, and **effective width,** BS. You must enter these as well as the actual **depth at discharge,** HD as shown in Figure 5.9. It takes some judgment to determine a characteristic rectangular shape, especially if there are large variations with shallow regions that may not affect the plume. This is partially due to the fact that CORMIX will only allow a variation of the actual discharge depth of ±30% of the average depth. The discharge may be raised above the bed, and HD is the bed depth at discharge, not the port depth. You will specify

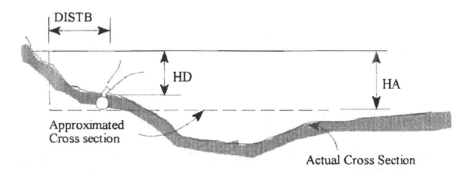

FIGURE 5.10   Unbounded receiving water simulated by CORMIX.

the port height above the bottom later. The variable DISTB is the **distance from the edge** of the simulated nearest rectangular shore to the discharge. For diffusers, this is to the center of the diffuser section. You also need to enter whether the channel is fairly **straight, moderately windy, or highly irregular**. This latter information is used to pick the correct far field calculations.

Figure 5.10 shows the representation with unbounded bodies where the plume will not reach the far shore. In this case you also specify the average depth, actual discharge depth, and the distance to the nearest bank. The width, however, is not entered. The average depth should be representative of the depth where the plume is expected to go. Don't worry about depths far from the discharge that will have not impact on the plume.

You next enter whether currents are **steady** or **tidal varying**. If steady, you need to enter the **average current**. Only one value is needed. CORMIX does not consider vertically varying currents. If flow is tidal varying, you can specify the **period**. The default value is 12.4 hours. You then need to enter the **instantaneous current** when predictions are to be made, the **time relative to slack tide** (zero current) when the predictions are to be made, and the **maximum tidal velocity** during the tidal cycle. CORMIX will calculate the rate of change of velocity and the instantaneous volumetric flow in the ambient. You need to enter either **Manning's n** or the **Darcy-Weisbach friction factor** at the bottom and the **wind speed** at the surface. Manning's n can vary from 0.015 for smooth bottoms to 0.2 for rocky channels with lots of weeds. A good wind speed is about 2 m/s for conservative predictions.

You next enter whether the ambient is **fresh** or **saltwater** and if it is *uniform* or **stratified**. If freshwater, you can enter the density directly or temperature and have CORMIX calculate the density. For salt water, you must enter the density. CORMIX has no utility to determine density from both temperature and salinity.* CORMIX represents ambient stratification with one of three approximations, as shown in Figure 5.11. You must select the one that most closely represents the stratification in the receiving body you want to consider. For type "A" with linear stratification, you will have to enter the **ambient density** at the **top** and **bottom**.

---

* The DENSITY program mentioned above in the RSB section can be downloaded from the Web from www.engr.orst.edu/~davisl.

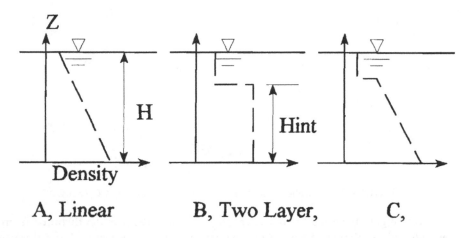

FIGURE 5.11   Ambient stratification types considered by CORMIX.

For type "B" with a step change in density, you also have to enter the **height of the pycnocline** (sudden change in density), HINT. For type "C" with a linear variation below the step, **the density at the top of the linear variation** is needed. If the actual stratification cannot be represented by any of these, select one that is closest to the density variation just above the outfall and below the pycnocline. Plumes will rarely go above the pycnocline, and density variations below it control plume behavior.

5.4.1.3   Discharge Data

CORMIX discharge data depend on which form of CORMIX you select, i.e., 1 for single port discharge, 2 for multiple port diffusers, or 3 for surface discharge. Each requires slightly different input. For **CORMIX1**, you are required to enter **which bank is closest,** the distance to the discharge port from that bank, the vertical and horizontal discharge angles, **theta,** and **sigma,** as shown in Figures 5.12 and 5.13, the **port diameter** (or area), and the **port height** above the bottom.

The input for **CORMIX2** is considerably more complicated because of all the possible arrangements for multiport diffusers. Referring to Figures 5.14 and 5.15, you are required to input the distance from the nearest bank to the **first port,** YB1, the distance from the nearest bank to the **end port,** YB2, and the **diffuser length,** LD. Note that the distance to the last diffuser port must be equal to the distance to the first port plus the diffuser length times the cosine of the diffuser orientation **angle relative to the current,** γ. If not, an inconsistency error will occur. The angles θ and σ are defined the same as before, but β and γ are also required. Beta is the port horizontal discharge direction relative to the diffuser line. Note that sigma plus beta must equal gamma or you get an inconsistency error and all three are required. You also need to enter the **port diameter, total number of ports** or risers, the number of ports on each riser, and the **port height** above the bottom, H_o.

You must also select the **type of diffuser** such as unidirectional, staged, alternating, or fanned, as shown on Figures 5.16, 5.17, and 5.18. Note that although CORMIX requires the selection of the diffuser type, it reduces all to an equivalent

PLAN VIEW

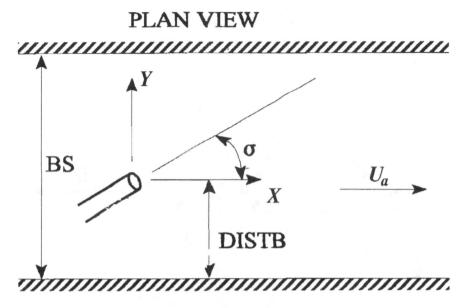

FIGURE 5.12   Plan view sketch of CORMIX1 discharge.

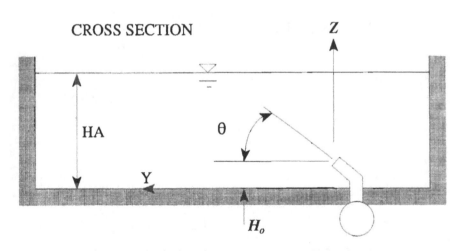

FIGURE 5.13   Elevation view sketch of CORMIX1 discharge.

slot diffuser with the exception of unidirectional diffusers with no control where the integral model CORJET is used in the near field. CORJET will be discussed in Section 5.4.3. It uses different hydraulic equations for the different types of diffusers but details near the diffuser are lost through the equivalent slot approximation. The **contraction ratio** allows you to account for the type of discharge nozzle. For nozzles

PLAN VIEW

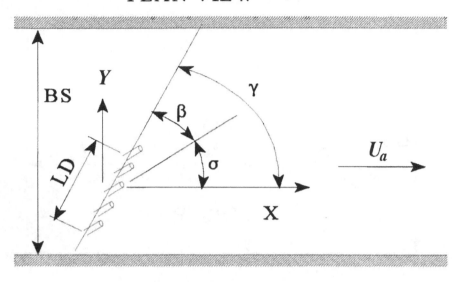

NEAREST BANK

FIGURE 5.14 Plan view sketch of CORMIX2 discharge with different angles required.

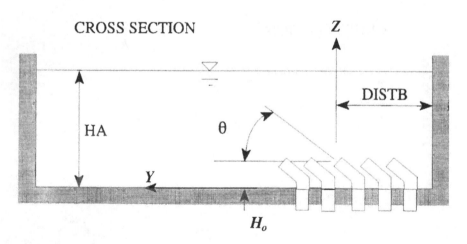

FIGURE 5.15 Elevation view sketch of CORMIX2 discharge.

with a rounded approach, a contraction ratio of 1.0 is appropriate. For orifices with square entrances, a contraction ratio of about 0.6 should be used to account for the vena contracta.

CORMIX3 input is slightly different, since it applies to surface discharge. Figure 5.19 shows a plan and elevation view of the type of discharge considered in CORMIX3. Note that in addition to the discharge **channel width,** β_o, and **depth,**

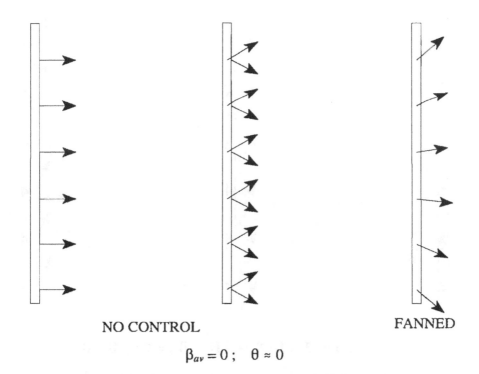

NO CONTROL   FANNED

$$\beta_{av} = 0 \; ; \quad \theta \approx 0$$

FIGURE 5.16 Diffuser types considered in CORMIX2 — Set 1.

H_o, the **ambient depth**, HDo, and **bottom slope**, θ, on the discharge bank are required. These latter two are required to see if the plume attaches to the bottom. If discharge is from a partially full pipe or other irregular-shaped discharge, the depth and width of an equivalent rectangular discharge should be entered. If discharge is from a pipe that is full, its **diameter** and invert depth are entered. The **discharge angle**, σ, is 0° for co-flowing discharge and 90° for perpendicular discharge. The exit geometry can be either **flush, protruding,** or **co-flowing,** as shown in Figure 5.20. If discharge channel is protruding, you will need to enter the **protrusion distance,** Y_o.

5.4.1.4 Effluent Data

Once the diffuser or channel data have been entered, you will be prompted for effluent information. This includes discharge **total flow rate** or exit **jet velocity,** and the effluent **density** or **temperature**. If the effluent is heated, you will have to enter a surface **heat transfer coefficient** in W/m²-K. The heat transfer coefficient has little effect on temperature in the near field but becomes important in the far field and for surface discharges. Table 5.2 gives some suggested values as a function of excess surface temperature and wind speed.[25] The discharge excess **concentration** above the ambient and **units** of the substance of concern are next. If temperature is of concern, you can enter the discharge excess temperature in degrees Celsius. If

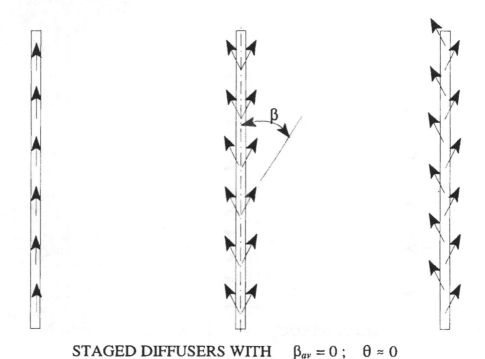

STAGED DIFFUSERS WITH $\beta_{av} = 0$; $\theta \approx 0$

FIGURE 5.17   Diffuser types considered in CORMIX2 — Set 2.

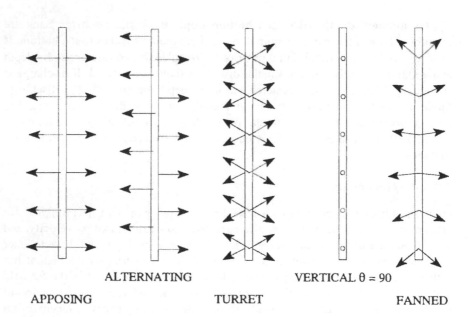

ALTERNATING   VERTICAL $\theta = 90$

APPOSING   TURRET   FANNED

FIGURE 5.18   Diffuser types considered in CORMIX2 — Set 3.

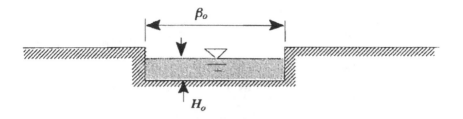

a) Discharge channel configuration

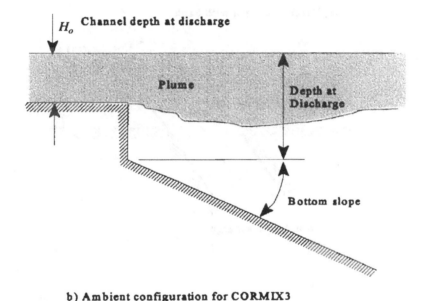

b) Ambient configuration for CORMIX3

FIGURE 5.19 Surface discharge channel geometry for CORMIX3.

the substance is not **conservative,** you will have to enter the **decay rate coefficient** in units/day. This concludes the input for discharge data.

5.4.1.5 Mixing Zone Data

The final set of data pertains to regulations. You will be asked if the effluent is **toxic or not** and if the EPA's toxic discharge regulations apply as given in Chapter 1. If it is toxic, you enter the **CMC** and the **CCC**. If there is a **water quality standard** of a conventional, nontoxic pollutant, you will need the value of the standard. If there is a **regulatory mixing zone,** RMZ, you will need to enter its extent as either a length, width, or area depending on the regulation. You can also enter the **region**

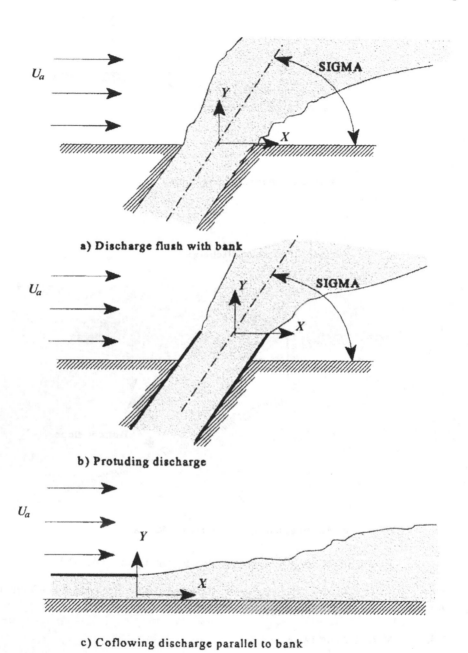

FIGURE 5.20   Types of exit geometry for CORMIX3.

of interest, ROI, which is the extent you want CORMIX to make its predictions.
Within the region of interest, you can specify the output interval. This specifies how
many output values you want printed out within the ROI and can vary from 3 to 50
values. Accuracy is the same regardless of the number of output values.

TABLE 5.2
Surface Heat Exchange Coefficients (W/m²-K) for Water
Surface With Small (0–3°C) Excess Temperature

Water Temp	Wind Speed (m/s)				
(°C)	0	1	2	4	8
5	5	10	14	24	42
10	5	11	16	27	49
15	5	12	18	31	59
20	5	14	21	38	68
25	6	16	25	45	82
30	6	19	30	54	100

Source: Jirka, G. H., Doneker, R. L., and Hinton, S. W., User's manual for CORMIX: A hydrodynamic mixing zone model and decision support system for pollutant discharges into surface waters, USEPA, 1996. Available on the INTERNET at ftp://ftp.epa.gov/epa_ceam/wwwhtml/cormix.htm.

5.4.2   CORMIX Hydraulics

5.4.2.1   Length Scales

The length scales used in CORMIX can be related to those proposed for RSB using the values of L, m, b, q, U_j, U_a, and N. Length scales that apply to both CORMIX1 and CORMIX2 are as follows:

$L_M = L^{1/4}m^{3/4}/b^{1/2}$ is related to the distance where flow changes from jet-like to plume-like in buoyant jet discharges.

$L_m = (Lm)^{1/2}/U_a$ is related to the distance a jet-like discharge penetrates into a cross flow before it is bent over and becomes a strongly deflected flow.

$L_b = Lb/U_a^3$ is related to the distance a plume-like discharge penetrates into a cross flow before it is bent over and becomes a strongly deflected flow.

$L_m' = (Lm)^{1/4}/N^{1/2}$ is related to the distance a jet goes before becoming strongly affected by stratification in a stagnant ambient leading to trapping.

$L_b' = (Lb)^{1/4}/N^{3/4}$ is related to the distance a plume-like flow goes before becoming strongly affected by stratification in a stagnant ambient leading to trapping.

Length scales that only apply to CORMIX2 with the **equivalent slot approximation** are as follows:

$\ell_M = m/b^{2/3}$ is related to the distance slot-type buoyant discharge changes from jet to plume behavior.

$\ell_m = m/U_a^2$ is related to the distance jet-type slot discharge penetrates a cross flow before becoming strongly bent over and advected by the ambient.

$\ell_m' = m^{1/3}/N^{2/3}$ is related to the distance a slot jet goes before becoming strongly affected by stratification in a stagnant ambient leading to trapping.

$\ell_a = U_a/N$ is related to the distance up or down that a plume travels before becoming strongly affected by a cross flow.

Length scales that only apply to **surface discharge** and CORMIX3 are as follows:

$L_M = (U_j Q)^{3/4}/(g'U)^{1/2}$ is related to the distance the initial jet region extends before changing into an unsteady surface layer.

$L_m = (U_j Q)^{1/2}/U_a$ is related to the distance the initial jet extends into a cross flow before being strongly deflected.

$L_b = (g'U_j)/U_a^3$ is a measure of the tendency of the surface spreading plume to have upstream intrusion.

In addition to determining the flow class for a particular discharge, these length scales are also used to determine the extent to which a particular hydraulic model or empirical expression is used. For example, when the predicted penetration for a particular crossflow single port discharge is less than L_m into the current, the jet-type strong current interaction module is used. When the distance reaches L_m, it switches to the strongly bent over module. This sometimes produces discontinuities in the dilution vs. distance curve produced because of the different diffusion rates.

The flow classes determined from these length scales are as follows:

CORMIX1	35 classes
Classes S:	Flows trapped in a layer within linear stratification.
Classes V, H:	Positively buoyant flows in a uniform density layer. Vertical or horizontal.
Classes NV, NH:	Negatively buoyant flows in uniform density layer. Vertical or horizontal.
Classes A:	Flows affected by dynamic bottom attachment.
CORMIX2	31 flow classes
Classes MS:	Flows trapped in a layer within linear ambient stratification.
Classes MU:	Positively buoyant flows in a uniform density layer.
Classes MNU:	Negatively buoyant flows in a uniform density layer.
CORMIX3	9 flow classes
Classes FJ:	Free jet flows without near-field shoreline interaction.
Classes SA:	Shoreline-attached discharges in a cross flow.

There are many subclasses within each class. For example, class MU8 has the following description, which you get when answering "yes" to the prompt for more information on this flow class:

An alternating multiport diffuser with predominantly perpendicular alignment is discharging into an ambient flow. For this diffuser configuration, the net horizontal momentum flux is zero so that no significant diffuser-induced currents are produced in the water body. However, the local effect of the discharge momentum flux is strong in relation to the layer depth and in relation to the stabilizing effect of the discharge buoyancy, so that the discharge configuration is hydrodynamically unstable. — The destabilizing effect of the discharge jets produces an unstable nearfield zone. For stagnant or weak cross-flow conditions, a vertical recirculation zone is produced leading to mixing over the full layer depth: however, the flow tends to re-stratify outside this zone that extends a few layer depths around the diffuser line. For strong cross-flow, additional stratification and mixing are produced.—

It is instructive to read the description of each flow class you get when first learning to use CORMIX. It will help you to understand what types of flows are generated under different conditions and how CORMIX handles the predictions. Experienced users can bypass much of the tedious details by selecting the FAST CORMIX option from the main menu.

5.4.3   CORJET

As was mentioned earlier, CORMIX uses the length scales to determine which hydraulic models to use in its predictions. If the diffuser is a line diffuser with unidirectional ports in a stable ambient, it makes its near field calculations using a three-dimensional integral model very similar to UDKHG called CORJET, coming from CORnel JET mixing zone model.[29] Rather than go through the complete theoretical background on this model, only the main differences between UDKHG and CORJET will be mentioned here.

1. CORJET uses Gaussian distributions rather than the 3/2 power law profiles used in UDKHG. The differences between these two distribution approximations was discussed in Chapter 1.
2. As a submodel to CORMIX, CORJET is limited to the ambient conditions given earlier, i.e., one ambient current and one of three types of stratification. It can be run, however, as a stand-alone model where input is generated from a separate file. In this mode, CORJET will consider not only several values of velocity and density as a function of depth but it will also allow the velocity to have different directions with depth.
3. CORJET does not make detailed calculations in the Zone of Flow Establishment as UDKHG does but approximates it in a single step, calculating the length of the zone of initial dilution from

$$L_e = D \left(6.2 - 20 \frac{\sqrt{1 - \cos^2 \theta \cos^2 \sigma}}{R} \right) R$$

Multiport diffuser - Corjet vs UDKHG

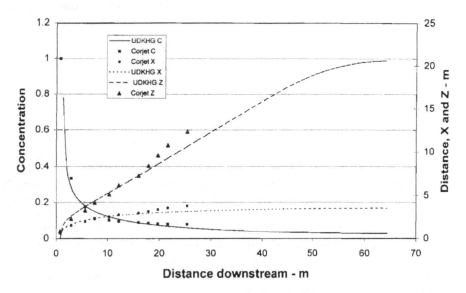

FIGURE 5.21   Plot of CORJET and UDKHG predictions for sample case.

with a flow angle relative to the x axis at this point given by

$$\gamma_e = \gamma_o \left(1 - \frac{1.22 \sin \gamma_o}{R} \right); \qquad R \geq 2$$

where R is the ratio of jet to ambient velocity and γ_o is the initial discharge total angle.

4. Merging is not gradual, but in one step when the plume characteristic diameter as determined from the Gaussian profile equals the port spacing. Calculations from this point on are based on a two-dimensional line plume.

5. CORJET allows you to select air as a working fluid and to calculate the density of air from its temperature. In UDKHG you must calculate the density of air yourself and enter density values directly.

Figure 5.21 is a graph showing the predictions of CORJET and UDKHG for a multiple port discharge into a flowing, stratified crosscurrent. It shows concentration, distance across the current X and the vertical distance Z as a function of distance downstream. As you can see, they agree well with each other for this case. Input values used in the calculations were as follows:

Discharge rate, $Q = 6.48$ m³/s
Discharge temperature and salinity = 30°C and 0.00
Diffuser: 11 ports spaced 1.82 m apart, 0.5 m diameter, in 30 m of water

```
#CORJET INPUT FILE
#Title line (50 characters max.):
Case5: MULTIPORT DIFFUSER: STRATIFIED, VARIABLE CURRENT
#Fluid (1=water,2=air), Density option (1=calculate,2=specify directly):
#Fluid (1/2):  Density option (1/2):  Ambient levels (1-10):
 1              1              3
#Ambient conditions (if d.o.=1, fill in TA+SA; if 2, fill in RHOA):
#Level ZA   TA    SA    RHOA  UA    TAUA
1      0.   12.   30.         0.5   0.
2      5.   15.   29.5        0.8   0.
3      15.  20.   28.         1.2   0.
#Discharge conditions (T0+S0, or RHO0 as above; if NOPEN=1: set LD=0,ALIGN=0):
#NOPEN D0  H0   U0    THETA0 SIGMA0 C0    KD   T0   S0   RHO0  LD   ALIGN
  11   0.5 0.   3.0   45.    45.    100.  0.   30.  0.         20.  60.
#Program control:
#ZMAX  ZMIN  DISMAX  NPRINT
 30.    0.    200.     10
```

FIGURE 5.22   Sample CORJET input file for stand-alone operation.

Vertical angle = 45° and horizontal angle = 45°
Ambient density and velocity at surface were 1.01947 gm/cm^3 and 1.2 m/s
Ambient density and velocity at bottom of 1.0227 gm/cm^3 and 0.5 m/s

UDKHG centerline concentrations were determined by taking the inverse of the centerline dilution. The centerline dilution was taken as 1/1.92 times the flux average dilution.

CORJET is easiest to run within CORMIX. It is automatically selected if it applies and the predictions appear in the standard CORMIX output. It can also be selected as a postprocessing option for those cases where it applies. In this case, the output is saved in CORMIX\POST\CJ\fn.CJX where fn is the file name you gave at the beginning of the session. If the flow is unstable as a result of bottom attachment, CORJET does not apply and will not provide predictions. If you elect to run CORJET as a stand-alone model, input is from an ASCII text file having any file name and extension with the data in five data blocks in free format (values separated by spaces). Each block is preceded by two dummy lines that are not read by the program. Figure 5.22 is a sample CORJET input file with the dummy lines indicated by a # sign. The dummy lines in this case have been used to clarify what the input values are.

The data required in the five blocks are as follows:

Block 1: A title line of your choice (50 character max)

Block 2: Option flags

First: 1 for water, 2 for air

Second: 1 for temperature/salinity input, 2 for density

Third: Number of ambient levels (maximum of 10)

Block 3: Ambient table having the number of lines given in option flags. Each
having level number starting at surface.

ZA:	Depth of ambient level
TA/SA/RHOA:	Temperature and salinity OR density at this level depending on value of second option flag
UA:	Ambient velocity at this level
TAUA:	Angle of ambient velocity vector measured CCW from x-axis, which is defined by the angle SIGMA0 given below

Block 4: Discharge conditions as follows:

NOPEN:	Number of ports: 1.0 for single port, 3 or greater for a multiple port diffuser
D0:	Port diameter
H0	Port height above bottom
U0	Jet velocity
THETA0:	Vertical discharge angle (90° is vertical)
SIGMA0:	Horizontal discharge angle measured CCW from x-axis (0° is co-flow, 90° or 180° is cross flow)
C0:	Discharge concentration (any units you chose)
KD:	Decay coefficient given as value/second (negative for growth)
T0/S0/RHO0:	Temperature and salinity or density depending on option flag 2
LD:	Diffuser length (set to zero if single port)
ALIGN:	Alignment angle of diffuser measured CCW from x-axis (0° for parallel and 90° for perpendicular)

Block 5: Output control given by:

ZMAX:	Maximum vertical distance, Z, from discharge of interest.
ZMIN:	Lowest value of Z of interest
DISMAX:	Maximum distance along trajectory of interest
NPRINT:	Number of print intervals (usually 5 to 20 is enough); this does not effect accuracy, just how much output you get

5.4.4 CORMIX Output

CORMIX output is generated in several forms. You will have the option to view and print out each. The files saved are the SIM\fn.CXn and SIM\CXn\fn.CXC files discussed above. Output variables are all abbreviated and are often confusing. Appendix 1 contains a list of CORMIX variables and their definitions. You may have to refer to this occasionally to understand what is printed out until you are familiar with the output. Also be aware that the dilution values given are sometimes centerline values and sometimes bulk average values. Be sure to read the headings. In addition, plume size is measured from a different reference once it attaches to a bank or shore.

5.4.4.1 Summary Output

You are first given the option to view or print a summary of the run. This summary is not saved in the form you view. So if you want a copy, you have to print it out. This does not work if you only have a network printer unless you have your computer configured to capture both network and local print jobs. The summary report is nice since it defines all the terms and gives the meat of the predictions. Even the summary report is quite wordy, so I'll only give a brief description of the format. The format is in blocks as follows:

- An echo of input variables.
- Values of flux variables for flow, momentum, and buoyancy.
- Length scales calculated from input.
- Dimensionless port Froude number, equivalent slot Froude number, and jet to ambient velocity ratio.
- Mixing zone/toxic conditions/and region of interest given in input.
- Flow class and description and applicable layer depth.
- Conditions at the edge of the near field as determined by the program. This is the field where mixing is rapid and may or may not apply to your problem. It gives concentration, dilution, location, and plume size at this point. These are equivalent to the values given by RSB.
- If this is a tidal unsteady case, the extent of predictions is given and statements regarding re-entrainment are made.
- It then gives an assessment of the effects of buoyancy, stratification, and instability.
- It gives the point where the plume becomes fully mixed with the ambient in a bounded environment such as a river.
- If the effluent is toxic and values of the CMC and CCC were given, it evaluates the discharge's ability to meet the EPA's toxic discharge conditions as given in Chapter 1. It checks each one and indicates whether they are met.
- If a regulatory mixing zone, RMZ, was given, the program gives the concentration, dilution, and plume size at the edge of the RMZ and

indicates whether the regulation given for this zone is met or not. If it was met within the zone, it indicates where.
* It then gives a series of comments on the design.

5.4.4.2 Full Prediction Report

After the summary report is viewed/printed, you are given the option to view/print the full prediction report. This is the report saved in the SIM\fn.CXn file. It can be accessed later and viewed or printed using any editor or word processor. This report contains all the information given in the summary report as well as all the intermediate predictions. The number of output prediction lines printed out is dictated by the output frequency you specified in the input. Unfortunately, many of the variables printed out in this complete file are just the abbreviations and you have to know what they mean.

After echoing the input variables, CORMIX prints out its predictions in *modules* with the output from one module becoming the input for the next. Usually predictions are continuous from one module to the next, but sometimes you notice a discontinuity, especially in plume size. The module number refers to which hydraulic model CORMIX is using for that module. The type of calculation is given after each module number along with the definition of any specific variables used, for example, whether it is giving centerline or average dilution and how plume size is measured. As predictions continue, the points where the CMC, CCC, and RMZ occur are printed out. In addition, it contains much information pertaining to what is happening as the plume develops, such as attachment to shore, surface, or bottom, trapping, upstream intrusion, instabilities, recirculation, etc. The last part of the report contains comments and suggestions on the design.

5.4.5 CORMIX Graphics

CMXGRAPH is a graphics package that is built into CORMIX that uses the output file to create various plots. You can either run CMXGRAPH at the end of a run or later as a postprocessing option. I find it best to use the postprocessing option after the output file is saved. That way you don't run a chance of corrupting the output file and losing it. Either way, when you select graphics, you will be prompted for the type of CORMIX the file was generated in and the name of the file. The graphics program is then executed using the output file for data. The first plot is usually a plan view of the whole plume with an option screen at the bottom. Figure 5.23 is a sample of a graph when the near field, side view is selected. This is for a multiple port line diffuser where CORJET is used in the near field and shows the jump you may get when going from one module to the next, in this case from CORJET to an empirical length scale model for a trapped terminal layer. This is a little deceiving, because CORJET uses the Gaussian (37%) characteristic width and the terminal layer module uses the full width. If the full 3/2 power law width were used in CORJET as in UDKHG, the discontinuity would be much less pronounced.

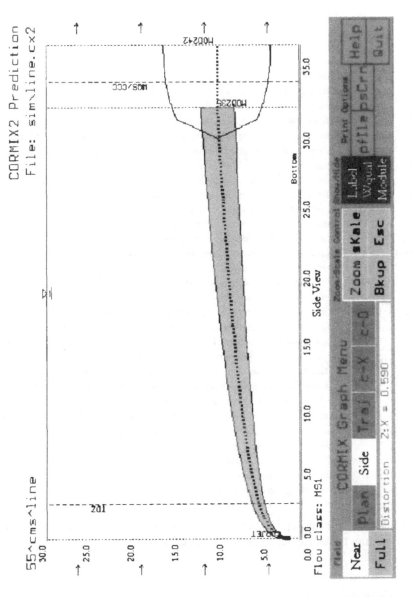

FIGURE 5.23 Sample CORMIX graphic screen showing side view with option buttons.

TABLE 5.3
Description of Option Keys in CORMIX Graphs

Variable	Key	Action
Near	N	Shows only the near field.
Full	F	Shows the full plume including near and far field predictions.
Plan	P	Gives the plan view.
Side	S	Gives the side (elevation) view.
Traj	T	Shows the X-Z trajectory.
c-X	X	Plots concentration versus distance in X direction.
c-D	D	Plots concentration versus distance along plume centerline.
Zoom	Z	Allows you to zoom in on a desired section of the graph. When you select zoom, move the curser to the upper left corner of the area you want and hit Enter, then move to the lower right corner and hit Enter again. CORMIX will zoom in on that area.
Bkup	B	Goes back to the previous screen. Similar to the WWW Back button.
sKale	K	Allows you to change the scale between axes. The present scale is shown as Distortion at the bottom of the option screen.
Esc	E	Cancels Zoom or sKale commands and reverts to default.
Label	L	Adds or removes project labels at top of plot.
Wqual	W	Adds or removes lines and labels on water quality standards such as where the CMC and CCC are met and where the RMZ is.
Module	M	Adds or removes lines and labels where CORMIX switches from one module to the next.
pflle	I	Prints a copy of the graph to a postscript file that can be printed out later. The file will have the project file name with an extension starting with P. The next letter in the extension will be either P, S, T, X, or D, depending on the view. The last letter in the extension will be a number from 0 to 9 that is automatically given by CORMIX. You must have software capable of printing a postscript file such as Ghostscript to print this graph. If you have a PostScript printer, you can simply copy the file to the printer and it will print it out as a graph.
psCrn	C	Removes the option menu from the bottom so you only have the graph. You can then use Shift-PrintScreen to print directly to your graphics printer or paste it to a Windows Paint utility as was discussed with UDKHG. Figure 5.24 is a sample plot of near field concentration vs. centerline distance using this method.
Help	H	Gives a help screen similar to this table.
Quit	Q	Leaves the graphics program and returns to the CORMIX menu.

Each option in the option screen can be executed by simply pressing a key corresponding to the capital letter in the option variable. An explanation of each is given in Table 5.3.

5.5 CORMIX EXAMPLES

Example 5.3

Treated industrial waste is to be discharged into a coastal environment. An LC50 bioassay shows that an average dilution of 1000:1 is required to prevent harmful

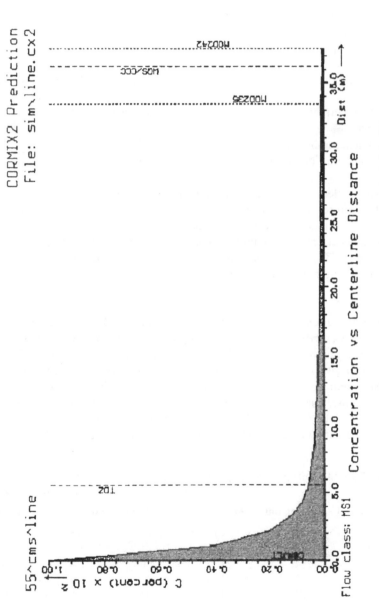

FIGURE 5.24   Sample concentration versus distance plot from CMXGRAPH.

effects on local aquatic life. The state where the outfall is to be built has put a limit of 150 m on the mixing zone where this dilution must be met and requires that CORMIX be used to determine dilutions. Conditions known are as follows:

Maximum peak dry weather flow	$= 0.6 \text{ m}^3/\text{s}$
Effluent density	$= 999 \text{ kg/m}^3$
Average depth where diffuser can be built	$= 95 \text{ m}$
Depth in center of region	$= 96 \text{ m}$
Worst stratification	$= \text{Linear } 1021 \text{ kg/m}^3 \text{ at surface} $ $1023 \text{ kg/m}^3 \text{ on bottom}$
Ambient current	$= \text{Varies in direction}$
Lowest 10%	$= 0.04 \text{ m/s}$
Maximum/minimum jet velocity	$= 3.0/0.5 \text{ m/s}$
Minimum port diameter	$= 2 \text{ in.}$

Your problem is to design a diffuser that will satisfy regulations. You should make the diffuser as short as possible with the fewest number of ports so that it doesn't cost too much. You can assume that the diffuser can be aligned perpendicular to prevailing current, that the ports can be about 1 m off the bottom, that the diffuser is far enough off shore that the plume will not become shore-attached, and that the pollutant is conservative (no decay).

Solution:

Since the currents vary in direction, we should consider either alternating ports or a diffuser with risers and two apposing ports. We can use the port velocity limits to determine total port area. Let's look at both discharge velocity limits. The higher velocity will give us the least number of ports and the shortest diffuser. To start, let's space them about 20 port diameters apart. This will give us good initial dilution before merging. Once we come up with a design, we can put them closer and try again to see if the resulting diffuser would be cheaper. With these constraints and using an alternating arrangement, the following table can be created:

Some Possible Diffuser Arrangements within Constraints

Diameter (in.)	Spacing (m)	Jet velocity = 3 m/s		Jet velocity = 0.5 m/s	
		No. of ports	Length (m)	No. of ports	Length (m)
2	1	99	100	592	600
3	1.5	44	67	263	400
4	2	25	50	148	300
5	2.5	16	40	95	240
6	3	11	33	66	200

For risers with two apposing ports, the length will be halved. Obviously, the 11 6-in. port diffuser that is 33 m long would be the cheapest alternating design. Let's start there and see if it works. If not, we can try the smaller diameters with the corresponding longer diffuser length. Running CORMIX, I had to make some assumptions. For example, I assumed an average Manning's *n* of 0.01 and no wind. These should have no effect in the near field. Right away I find some limits in CORMIX. It will not let me use a diffuser that is shorter than the water depth. For the high velocity case, this means I can only simulate the 2-in. port case with CORMIX.

Reenter the data for the 98 2-in. case with a 100 m diffuser length. The only question you may not have an answer for is the contraction coefficient. Let's assume a rounded nozzle with a coefficient of 1.0 to reduce head losses. For a distance to the first port, select some large number such as 3000 so the plume won't have shore attachment. The distance to the last port then becomes 3100 m, since the diffuser is perpendicular to the shore. When CORMIX runs, view the summary report. Under Regulatory Mixing Zone Summary, you find

Dilution = 202
X = 208.72

..."the specified ambient water quality standard was not encountered within the plume region..." This means that at 208 m downstream, the dilution is only 202. We want 1000 at 150 m.

O.K., that one does not work. Let's try the longest diffuser, the 600 m diffuser with 592 2-in. port with low exit velocity. Work your way through the prompts answering NO or EXIT to everything until you get to the red ITERATION MENU that allows you to rerun with a different ambient, discharge, or regulation. Select *different discharge* and enter a new description and file name. Reenter the diffuser data, but this time, use 582 ports and 600 m for a length. Figure 5.25 contains sections of the summary report showing the input values I used. Figure 5.26 shows the regulatory mixing zone section of the summary report. As you can see, CORMIX indicates a dilution of 1418.7 at the edge of the mixing zone 150 m downstream. Notice also that CORMIX simulates the alternating port arrangement as vertical ports, since the net horizontal momentum is zero. This is o.k. for the far field, but tends to under-predict dilutions in the close proximity of the diffuser.

This is quite a bit more than we need, so lets look at the 6-in. ports at low discharge velocity. This gives a diffuser with 66 ports 200-m long. Work your way through CORMIX prompts until you get the iteration menu again. Reenter data to reflect the 200-m, 66-port diffuser and rerun. This time the summary report shows a dilution of 941.2 at the edge of the mixing zone. Not quite enough.

Making several other iterations, I found the following: A 300-m diffuser having 148 4-in. alternating ports produces a dilution of 1094 at the edge of the mixing zone. If I use the same length diffuser with 148 4-in. ports on risers with two apposing ports per riser, I get a dilution of 1052. Either of these two designs will satisfy the design constraints. This, however, is a case where a novice user may feel he has reached a good design, since CORMIX says it will work. It should be pointed out that low discharge velocities such as 0.5 m/s used in this example are not usually recommended unless it is for low, dry weather conditions. Appendix 2 contains the complete output prediction file for the 300 4-in. alternating port run.

```
SUMMARY OF INPUT DATA:

    AMBIENT PARAMETERS:
    Cross-section                                      = unbounded
    Average depth                        HA  =         95      m
    Depth at discharge                   HD  =         96      m
    Ambient velocity                     UA  =         .044    m/s
    Darcy-Weisbach friction factor       F   =         0.0017
    Calculated from Manning's n              =         .01
    Wind velocity                        UW  =         0       m/s
    Stratification Type                  STRCND         = A
    Surface density                      RHOAS      =   1021  kg/m^3
    Bottom density                       RHOAB      =   1023  kg/m^3

    DISCHARGE PARAMETERS:      Submerged Multiport Diffuser Discharge
      Diffuser type                      DITYPE  =
    alternating perpendicular
      Diffuser length                    LD   =    600 m
      Nearest bank                            =    left
      Diffuser endpoints                 YB1  3000 m;    YB2    3600 m
      Number of openings                 NOPEN =        592
      Spacing between risers/openings    SPAC  =        1.01   m
      Port/Nozzle diameter               DO   =         .051   m
        with contraction ratio              =         1
      Equivalent slot width              BO   =        0.0020  m
      Total area of openings             AO   =        0.0020 m^2
      Discharge velocity                 UO   =        0.49   m/s
      Total discharge flowrate           QO   =        .6     m^3/s
      Discharge port height              HO   =        1      m
      Nozzle arrangement                 BETYPE = alternating without fanning
      Diffuser alignment angle           GAMMA      =              90 deg
      Vertical discharge angle           THETA  =     90.0 deg
      Horizontal discharge angle         SIGMA  =     0.0 deg
      Relative orientation angle         BETA   =     90.0 deg
      Discharge density                  RHOO   =     999 kg/m-3
      Density difference                 DRHO   =     23.9791 kg/m^3
      Buoyant acceleration               GPO    =     .2299 m/s'2
      Discharge concentration            CO     =     1000 percent
      Surface heat exchange coeff.       KS     =     0 m/s
      Coefficient of decay               KD     =     0 / s
```

FIGURE 5.25   Section of CORMIX Summary Report showing input for Example 5.3 using a 600 m diffuser with 592 2-in. alternating ports.

Example 5.4*

A plant discharges 2.2 m³/s (50 mgd) of wastewater at 20°C with a copper content of 80 μg/l into a large estuary. Discharge is from a 2-m wide-open channel whose invert is 1.0 m below Mean Low Water (MLW). The estuary depth increases rapidly from shore until the bottom levels out at about 5 m below MLW. Figure 5.27 shows the tidal variation of velocity and elevation relative to MLW for the site in question. Using this figure, we see that the depth in the discharge channel varies

---

* This example is a slightly modified version of the transient example given in Appendix D of the *User's Manual for Cormix: A Hydrodynamic Mixing Zone Model and Decision Support System for Pollutant Discharges into Surface Waters*, by Jirka, G. H., Doneker, R. L., and Hinton, S. W., which can be downloaded from the Web with the program as mentioned previously.

```
*******************REGULATORY MIXING ZONE
SUMMARY**************************
The plume conditions at the boundary of the specified RMZ are as follows:
                Pollutant concentration              =        .704830 percent
                Corresponding dilution               =       1418.7
                Plume location:            X =        150.00  m
                (centerline coordinates)   Y =           .00 m
                                           Z =          7.22 m
                Plume dimensions:      half-width     =       762.97  m
                                       thickness      =        12.67  m
Furthermore, the specified water quality standard has indeed been met within the RMZ.  In
particular:
The ambient water quality standard was encountered at the following
                plume position:
                Water quality standard               =      1 percent
                Corresponding dilution               =      1000
                Plume location:            X =       30.14 m
                (centerline coordinates)   Y =          .00 m
                                           Z =         7.22 m
                Plume dimensions:      half-width     =      270.83 m
                                       thickness      =       24.77 m
```

FIGURE 5.26 Section of CORMIX Summary Report showing Regulatory Mixing Zone results for 600-m diffuser with 592 2-in. alternating ports.

from 0.5 to 1.8 m during one cycle. The ambient velocity varies from about −0.7 to +0.7 m/s during the cycle. The density of the ambient is approximately 1018 kg/m^3.

State regulations give a 250-m mixing zone in any direction from the outfall. The values for the CMC and CCC are 25 and 15 µg/l, respectively. The problem is to see if the regulations are satisfied at different times during the tidal cycle.

Solution:

As was mentioned previously, CORMIX is basically a steady-state model, but it can consider the effects of tidal changes at a particular time during the cycle. It is usually a good idea to make several simulations at critical times during the cycle such as hourly increments following slack tide when re-entrainment is important. It should be noted that CORMIX will not accept zero time and velocity (slack tide) as input. You should start an hour or half hour after slack tide where time and velocity are small but not zero. For this example, let's look at 1 h after slack tide (at about 11:45 on the tidal plot) where the ambient velocity is 0.22 m/s. The input values of a CORMIX3 simulation are listed below:

Water body depth	5.65 m
Depth at discharge	5.65 m
Unbounded	
Tidal period	12.4 h
Max tidal velocity	0.75 m/s

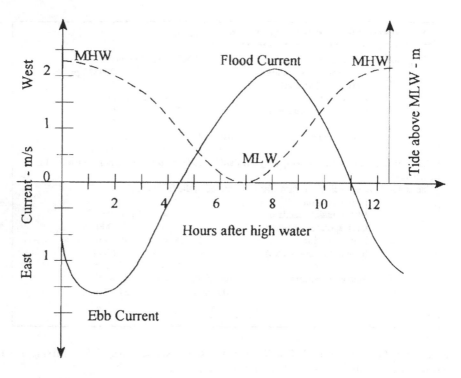

FIGURE 5.27　Velocity and elevation for one tidal period for Example 5.4.

Tidal period after slack tide	1.0 h
Tidal velocity at this time	0.22 m/s
Darcy–Weisbach f	0.025
Wind speed	2 m/s
Average ambient density	1018.0 kg/m³
Right bank	
Flush	
Horizontal angle sigma	90°
Depth at discharge	2.15 m
Bottom slope	11°
Width	2.0 m
Depth	0.65 m
Effluent flow rate	2.2 m³/s
Effluent fresh @	22°C
Not heated	
Effluent concentration	80 µg/l
Conservative	
Toxic	Yes
CMC	25
CCC	15

WQ standard of conventional pollutant	No
Regulatory mixing zone	distance = 250 m
Region of interest	2000 m

Enter the values and run CORMIX3. Notice that CORMIX stops when $x = 282$ m. This is because the time to get there has been too long for CORMIX to give a good accounting of tidal changes. From the toxic report, however, you will see that all three U.S. EPA conditions are satisfied. If the model had been run using steady-state conditions, ignoring tidal variations, the concentration at the edge of the mixing zone would have change from 18 μg/l to 10 μg/l, showing the effect of tidal reversal.

Problems

5.1 A toxic industrial waste is discharged at 31°C into a large river at a discharge rate of 0.002 m³/s through a single 5 cm port. If the river is 3-m deep, has a temperature of 21°C and an average velocity of 15 cm/s, determine the dilution at the EPA toxic zone of initial dilution conditions. Which is the most restrictive? Repeat the problem using UDKHG.

5.2 A 20-port line diffuser is to be used to discharge 0.3 m³/s of wastewater into a 5-m deep river. The ports are 5 cm diameter spaced 3 m apart on both sides of the diffuser. Discharge is horizontal and the diffuser is perpendicular to a 12 cm/s current. The density of the effluent and ambient are essentially the same. If the effluent is initially 10° warmer than the ambient, determine the average plume temperature 5 m downstream using CORMIX2.

5.3 Use CORMIX to solve Problem 3.3.

5.4 Use RSB to solve Problem 3.3.

5.5 30 mgd of treated industrial waste are discharged into a shallow river through a diffuser that has eight 8-in. ports spaced 2 m apart. Discharge is horizontal and 90° from the current (cross current). If the river is 2.44 m deep and has an average velocity of 0.18 m/s, use CORMIX2 to determine the dilution 10 and 100 m downstream. Ignore density differences.

5.6 Use CORMIX3 to solve Problem 4.3.

6 Atmospheric Discharges

6.1 INTRODUCTION

There are a large number of atmospheric models that have been developed to predict dispersion in the atmosphere including stacks, automobile, and agriculture pesticides emissions. Since this book emphasizes mixing zones, only point source emission models will be discussed. Although similar in many respects, atmospheric discharges may possess several complications that are not present in aquatic discharges. These differences are mainly due to moisture either as condensed water vapor or from drift from cooling towers. Drift consists of water droplets that are entrained in the discharge from cooling towers and carry with them the chemicals and dissolved solids present in the water being cooled. The condensed water vapor can cause a visible cloud that can reduce visibility to either aircraft or vehicles on the ground. In addition, it can shadow large areas, reducing exposure to the sun and reducing plant growth. If the cloud reaches the ground during freezing conditions, ice buildup on plants, roads, and buildings can be a problem. The other obvious difference between aquatic and atmospheric discharges is that with atmospheric discharge there is no water surface to limit plume rise. As a result, wind and atmospheric density variations play a dominant roll in atmospheric plume rise.

The thermal effects due to condensation and evaporation and the mass loss effects due to droplet deposition have little effect on the main properties of the plume such as trajectory, size, and entrainment of ambient air. As a result, the UDKHG, UM, and CORJET models discussed in previous chapters can be used to predict near field properties of atmospheric discharges. In fact, earlier versions of UDKHG and UM were specifically written to simulate cooling tower discharges.[30,31] CORJET has a switch that allows you to specify whether you want to simulate an aquatic or atmospheric discharge. In UDKHG and UM, enter the appropriate air density and temperature and leave the salinity blank. Use a very deep depth to simulate the atmosphere.

There are, however, many models that have been specifically developed for stack discharges without moisture effects. Most of these are so-called Gaussian plume or puff models where the spacial distribution within the plume or puff is assumed to be Gaussian and both the trajectory and size are calculated independently by different schemes. Several good models have been written specifically for cooling tower plumes that include moisture effects and plume visibility. Examples are Carhart and Policastro,[32] Winiarski and Frick,[31] Hanna,[33] Orville et al.,[34] Slawson and Wigley,[35] and Macduff and Davis.[30] Only the Carhart and Policastro model, known as the ANL/UI model, will be discussed in any detail in this book, because it does a good job of predicting both plume rise and visibility and it is the model used in the Seasonal/Annual Cooling Tower Impacts model[36-38] to be discussed in Section 6.4. In addition, a brief discussion of Gaussian plume methods will discussed.

6.2   GAUSSIAN PLUME AND PUFF MODELS

Several simplifying assumptions are made in Gaussian plume models that make them fairly simple. First, it is assumed that the plume or puff moves horizontally at the local wind speed. This is known as the "bend over plume assumption." It ignores the short region near the source where the discharge momentum resists bending over with the wind. As a result, these models are only good starting several discharge diameters downstream. The other assumption is that the spatial distribution within the plume is Gaussian. For example, for a continuous discharge from a source, the downstream concentration can be approximated by

$$C = \frac{\dot{M}}{U_a 2\pi R^2} e^{-\frac{r^2}{2R^2}} \tag{6.1}$$

where $\dot{M}$ is the rate of discharge (kg/s), R is the plume radius, and r is the radial coordinate perpendicular to the plume centerline. Thus, the centerline concentration decays as the inverse square of the plume radius, inversely with wind speed, and radially outward as a Gaussian function. Only the plume radius and height of rise remain to be found. For a single puff of material, Equation (6.1) becomes

$$C = \frac{M}{4\pi R^3} \sqrt{\frac{2}{\pi}} e^{-\frac{r^2}{2R^2}} \tag{6.2}$$

where M is the total mass in the original puff. If a series of puffs are considered and superimposed, Equation (6.2) can be reduced to Equation (6.1). Since each individual puff can have a different initial concentration, the Gaussian puff method can be extended to handle transient discharges.

One common expression for plume growth as a function of a turbulent dispersion coefficient, k, is given by

$$R = \sqrt{2k \frac{x}{U_a}} . \tag{6.3}$$

However, many people take advantage of the classic work by Turner[39] and Briggs[40] where the plume growth and rise were determined experimentally for a wide number of cases based on atmospheric stability. Briggs found that buoyant plumes follow the "⅔ power law" for a considerable distance downwind regardless of stratification. The empirical expression for plume rise in English units that fit the majority of cases was

$$h = 1.6F^{1/3} \frac{x^{2/3}}{U_a} \tag{6.4}$$

where h is the plume rise, x is the distance downwind, and F is a discharge buoyancy function given by

$$F = \frac{\Delta\rho}{\rho} g U_o R_o^2 \qquad (6.5)$$

The initial density difference between plume and ambient is $\Delta\rho/\rho$, U_o is the exit velocity, and R_o is the stack exit radius.

It becomes a simple matter to apply the plume size as determined by Turner to the Gaussian plume model which gives the concentration of a plume whose rise is determined by Briggs expressions. It should be noted that Equation (6.4) does not include stack wake effects and is not valid in the extreme far field. Refer to the original Briggs reference for more information.

6.3   THE ANL/UI COOLING TOWER PLUME MODEL

The ANL/UI model (Argonne National Laboratory/University of Illinois) is an Eulerian integral model similar to UDKHG, except it was designed specifically for cooling towers and includes moisture thermodynamics, drift deposition, tower wake effects, and ground shadowing. To more closely simulate experiments, the ANL/UI model uses different spreading rates for momentum, thermal energy, and moisture. This results in downstream sizes of $R = R_m/\sqrt{\nu}$ and $R_w = \sqrt{\lambda/\nu}\, R_m$ where R_m is the momentum plume size, R is the temperature plume size, and R_w is the moisture core size. The constants λ and ν are calibration constants. They were found to be approximately 0.51 and 1.2, respectively.

The equations used in the ANL/UI model, up to the atmospheric diffusion phase, are developed in detail by Carhart and Policastro.[32] They are briefly presented here to show what must be added to the equations presented in Chapter 4 to account for thermal and moisture effects. In these equations, U is the plume velocity, U_a is the wind speed, X is the specific humidity kg/kg-dry-air, σ is the liquid mixing ratio, kg/kg-dry-air, γ_d is the dry adiabatic lapse rate, h_{fg} is the latent heat of vaporization, C_p is the constant pressure specific heat, and Q_s is the saturation mixing ratio at temperature T and pressure p. The horizontal plume velocity was assumed to be equal to the ambient velocity. The equations are written with various fluxes, Φ, as dependent variables defined as follows with subscripts "p" referring to the plume, "a" to the ambient, "d" to dry air, and z as vertical:

Mass:

$$\Phi_m = \pi R_m^2 \bar{\rho}_p U \qquad (6.6)$$

Horizontal momentum:

$$\Phi_{hm} = \Phi_m U_a \cos\theta \qquad (6.7)$$

Vertical momentum:

$$\Phi_{vm} = \Phi_m U_z \qquad (6.8)$$

Enthalpy:

$$\Phi_h = \frac{C_p T_p + \lambda h_{fg} X_p}{v} \Phi_m \qquad (6.9)$$

Total water:

$$\Phi_w = \lambda \Phi_m \frac{X_p + \sigma}{v} \qquad (6.10)$$

Liquid water:

$$\Phi_{lw} = \frac{\lambda \sigma}{v} \Phi_m \qquad (6.11)$$

The fractional entrainment rate is given by

$$\mu = \frac{2\rho_a}{R_m \overline{\rho}_p U} \left(\alpha |U - U_a \cos\theta| + \beta U_a \sin\theta \right) \qquad (6.12)$$

The constants α and β are experimentally determined entrainment coefficients. With these fluxes and the entrainment function defined, the following system of ordinary differential equations can be developed.

Mass:

$$\frac{d\Phi_m}{ds} = \mu \Phi_m \qquad (6.13)$$

Horizontal momentum:

$$\frac{d\Phi_{hm}}{ds} = \mu \Phi_m U_a \qquad (6.14)$$

Vertical momentum with buoyancy:

$$\frac{d\Phi_{vm}}{ds} = \Phi_m \frac{g}{vU} \left(\frac{T_p^* - T_a^*}{T_a^*} - \lambda \sigma \right) \qquad (6.15)$$

Energy:

$$\frac{d\Phi_h}{ds} = -C_p \Phi_m \sin\theta \gamma_d \frac{\rho_a}{\rho_p} + \frac{\mu}{\nu}\left(C_p T_a + \lambda h_{fg} X_a\right)\Phi_m \qquad (6.16)$$

Total water:

$$\frac{d\Phi_w}{ds} = \mu\Phi_m \frac{\lambda}{\nu} X_a \qquad (6.17)$$

Liquid:

$$\frac{d\Phi_l}{ds} = \frac{C_p \Phi_m}{\nu h_{fg}} \frac{\chi}{\chi+1}\left\{ \left(\gamma_d \frac{\rho_a}{\rho_p} + \frac{\Pi}{\tau}\right)\frac{U_z}{U} + \mu\left[\left(T_p - T_a\right) - \frac{X_p - X_a}{\tau}\right]\right\} \qquad (6.18)$$

$$\frac{dx}{ds} = \frac{U_a}{U} \qquad (6.19)$$

$$\frac{dz}{ds} = \frac{U_z}{U} \qquad (6.20)$$

with

$$\tau = \frac{\partial Q_s}{\partial T} \qquad (6.21)$$

$$\chi = \lambda h_{hg} \frac{\tau}{C_p} \qquad (6.22)$$

$$\Pi = -\rho_a g \frac{\partial Q_s}{\partial p} \qquad (6.23)$$

These equations are sufficient to integrate step-wise along the trajectory of the plume to determine bulk plume properties. Plume visibility is determined by using bulk average plume temperatures and moisture content along with equivalent Gaussian profiles and saturation conditions. The assumed profiles as a function of radius are

$$T_p(r) = T_a + 2\left(T_p - T_a\right)e^{-\frac{r^2}{b^2}} \qquad (6.24)$$

and

$$X_p(r) = X_a + 2\left(X_p - X_a\right)e^{-\frac{r^2}{b^2}} \qquad (6.25)$$

where $b = R/\sqrt{2}$. Tower wake effects are included by imposing a drag force in the vertical momentum equation that draws the plume into the wake of the tower. Different empirical equations are used for natural draft and mechanical draft towers for wake drag.

The model was extensively tuned to both field and laboratory data from Europe[41,42] and the United States.[43-45] In subsequent verification runs, the model was found to predict plume rise within a factor of 2.0, the visible plume length by a factor of 2.5, and dilution within about 30%. This was considered good, since other models were generally not better, and worse in most cases. Input and output from the model is handled through the SACTI model described below.

6.4   SEASONAL/ANNUAL COOLING TOWER IMPACT MODEL, SACTI

All models presented to this point are usually used to predict "worst-case" conditions. In many atmospheric cases, average or accumulative behavior over a long period of time, such as sunshine reduction as a result of plume shadowing or drift deposition, is important. Visible plumes from cooling towers behave very much like clouds and reduce any solar radiation passing through them. To accomplish this, meteorological conditions at the selected site must be used to predict all possible plumes, determine the frequency of occurrence, and integrate the resulting shadowing effect and drift deposition. Since the wind usually varies in all directions such as shown on the wind rose in Figure 6.1, shadowing and drift deposition will vary around the discharge with higher values in direction of the prevailing wind. This results in contour lines around the discharge where accumulative values over a given period would occur.

Several models have been developed to predict seasonal averages. The most noted are the Swiss KUMULUS[46] model and the Seasonal/Annual Cooling Tower Impact model, SACTI,[38] developed by staff at the Argonne National Laboratory and the University of Illinois for the Electric Power Research Institute (EPRI).* The SACTI model was patterned after the KUMULUS model but has several improvements that make it predict actual behavior better. Since these models are much more complicated than others considered in this book and because they are not readily available on the Internet, only the SACTI model will be considered in this book and then only in brief form. Readers interested in this type of model should get the model and manual from Argonne National Laboratory or from EPRI.

The SATCI model can consider any number of mechanical or natural draft towers arranged in any configuration. It uses standard 5-year meteorological tapes for environmental input and assigns each hourly record to one of 864 bins. It lumps and categorizes these records until approximately 30–100 plume categories (user-specified) are determined. These categories become characteristic to the selected site with a frequency of occurrence associated to each. The plume models are then run for each category and the results summed and averaged to determine the seasonal/annual impact.

---

* The SATCI model and the imbedded ANL/UI plume model along with a complete user's manual and sample case can be obtained from the Argonne National Laboratory through Anthony J. Policastro, e-mail, policastro@anl.gov.

Annual Wind Rose for Syracuse, New York
Period of Record: 83–87

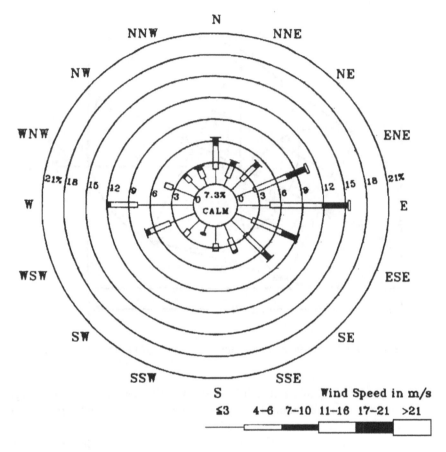

FIGURE 6.1 Typical wind rose with bars in the direction of the wind. Speed and frequency are given by the length and type of bar.

The SATCI code consists of several submodels with the output of one being the input to the next. The function of these submodels is as follows:

Preprocessor: The preprocessor reads the meteorological tape and uses the information on it and site specifics to perform the categorization.

Single plume module: The single plume model used in SATCI is the ANL/UI model discussed above.

Multiple plume module: The multiple plume submodel is an extension of the single plume model with a merging process that allows plumes from different locations to merge to form single elongated plumes. It follows each individual plume separately, starting with the upwind plumes first, until it merges with another one downwind. The merging process is similar to the

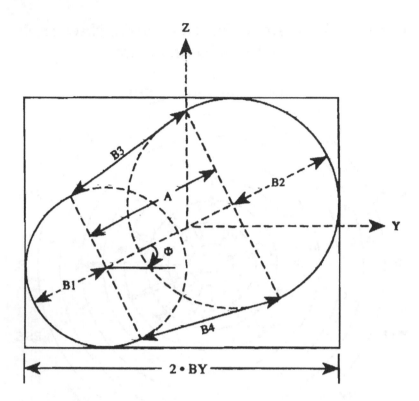

FIGURE 6.2   Merging of two plumes to form an elongated single plume.

one presented by Wu and Koh,[47] where two different sized plumes merge
into one elongated plume, as shown in Figure 6.2 with the form preserving
the basic cross-sectional shape and all fluxes.

Single plume drift model: The single plume drift model keeps track of the
water particles within the plume, when they break away and where they
end up on the ground. The drift droplet composition is specified by the user
giving the number of drop sizes, total drift rate, total dissolved solids
concentration in droplets, size and mass fraction of drops in each drop class.
The program determines the trajectory of each droplet class based on the
plume velocity and the droplet fall velocity, with each droplet class having
a different fall velocity. The mass of each drop may change as evaporation
or condensation takes place. If sufficient evaporation takes place before the
drop hits the surface, only the dry solid porous cap will remain. Once a
droplet or particle breaks away from the plume, its ballistic trajectory is
followed until it hits the ground, where it is summed with others to determine the total deposition. Details of the drift model can be found in Dunn,
Gavin and Boughton.[48]

Multiple plume drift: The multiple plume drift submodel uses the single plume
drift model for the upwind cell, then uses a Gaussian distribution with angle

to account for the displacement of deposition from cells other than the upwind one.

The shadowing submodel: Shadowing depends not only on the plume size and location but also on the time of day as a result of the solar angle. The shadowing submodel uses hourly calculations during each day to determine the hours of shadowing at a given location, the total beam solar energy, the percentage of total solar energy, and the percentage of beam energy.[37]

Examples of some of the plots that can be generated using the SACTI model are shown in Figures 6.3, 6.4, and 6.5.[38] Figure 6.3 shows the plume length frequency

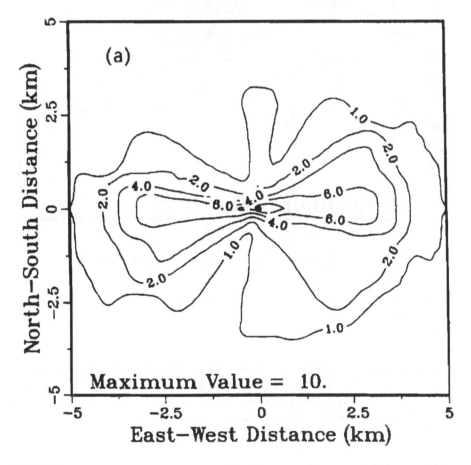

FIGURE 6.3 Typical plot showing plume length contours for a natural draft tower. Values represent percentage of occurrence.[38] (Used with permission.)

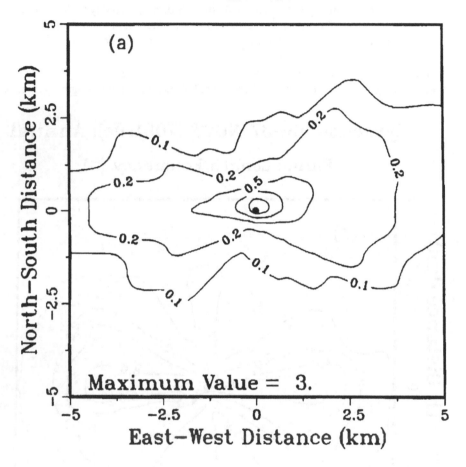

FIGURE 6.4 Typical plot showing total solar energy deposition loss in percentage.[38] (Used with permission.)

around a natural draft tower. The contour lines represent percentage of occupance. Figures 6.4 and 6.5 show the total solar deposition loss in percentage and salt deposition from drift in kilograms per square kilometer per month for the same cooling tower. They all show the effect of the prevailing wind in an east–west direction. Notice that salt deposition extends 10 km from the tower.

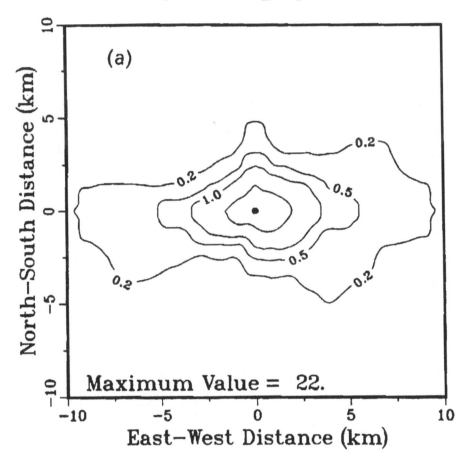

FIGURE 6.5 Typical plow showing drift salt deposition (kg/km²-month) for a natural draft
tower.[38] (Used with permission.)

7 Accuracy, Recommendations, and Closure

7.1 ACCURACY

The question is always asked, "Just how accurate are the predictions?" The answer is not easy. Plume prediction is not an exact science. No matter which method you use, assumptions and simplifications have to be made that are different for different types of flows. The results are models that give better simulations in some situations and less in others. None of them give exact answers. Let's look at some of the problems.

First, what are you comparing the predictions to? Field measurements themselves are not exact. Because of not knowing where the plume is, turbulent fluctuations, and equipment limitations, measured values may not be what you expect. I know of one instance where thousands of dollars were spent trying to get field measurements to verify a mathematical model. Helicopters flew over the visual part of the plume and lowered traps into the water to select samples at various locations and depths. However, divers who were sent down to photograph the plume found that a shear current caused a major portion of the plume to go in a different direction than was observed by the helicopters. The helicopters were taking measurements in the wrong place. Even if they had been in the right place, the traps would have taken samples of the water column at the instant when the trap closed. Due to large-scale eddies that exist in most plumes, the concentration at a particular point can be completely different at one instant compared to another. Mathematical models always give time-averaged values. Laboratory measurements also have problems. Wall effects, problems in turbulence scaling, and geometry scaling problems produce differences between actual plumes and simulated plumes.

The models have inherent restrictions due to their formulation. For example, the integral models, PDSG, UDKHG, UM, and CORJET ignore boundaries or include them in an approximate manner. The basic approach in integral models is to assume similar profiles as you march along. This means ambient water is assumed to exist on all sides. The profiles can be modified as they are when plumes merge, but a different model would have to be written for each site to include all boundary effects. As a result, if these models are allowed to continue to predict after the plume hits a boundary, the predictions are in error. The image method mentioned in Chapter 4 is useful to simulate shallow water discharges if used properly, but this an exception to the rule.

Empirical models such as CORMIX that use different hydraulic equations for different flow classes will have discontinuous predictions when going from one class to another. The predictions may be good when data fall in the middle of a class where the equations were generated, but deviate from reality when used to extrapolate to the limits of the class. For example, in a particular case where 2.7 m³/s were discharged into a large stratified river through 2.36-cm diameter ports spaced 6 m apart, CORMIX predicted a class MS7 with an initial dilution of 290 when 110 ports were used and a class MS5 with a dilution of 1050 when 120 ports were used. Only a very large number of classes would produce smooth transition from one class to another. All models have been "tuned" to experimental data. The experimental data used may have been biased toward a particular type of discharge or have other experimental errors resulting in poor or biased predictions by the models.

In order to give you an idea of the prediction capabilities of the different models, I have generated several plots comparing model predictions to each other and to data that can be found in the literature. I'm sure that if you were to use different data, you would get different results than I have shown here. You can find one study that shows one model doing much better than the others, while a different study shows that model to be much worse than the others. The figures I present here are meant to show relative differences between models.

Figure 7.1 shows the trajectory as predicted by UDKHG, UM, and CORMIX compared to the laboratory measurements of Kannberg and Davis[13] for two different ambient current rates. The discharge was from a line diffuser with vertical ports spaced 5 discharge diameters apart at a discharge densimetric Froude number of 10. The two currents were $R = 0.5$ and $R = 0.05$ where R is the ratio of current velocity to discharge velocity. The Kannberg and Davis data were obtained by tracing photographs of the dyed plume through transparent channel walls. Notice the scatter in measured data for repeated runs. This is typical of experimental scatter. For this case, UDKHG and CORMIX were slightly better then UM at low current. All three models did well at the higher current.

Figure 7.2 is a plot of average plume dilution as predicted by UDKHG and CORMIX versus downstream distance as compared to measurements in the field using a dye study in the Cook Inlet.[49] The diffuser in this case was turret type with a riser with three 9 cm ports oriented at 45°, 90°, and 135° from the current. The predictions were obtained by running the models for each port separately and then averaging the results. At the time of this study, the discharge was 1.1 m³/s and the ambient was 4.6 m deep with a 0.4 m/s current. The measured values were taken from a boat that followed a drifter in the current. Dye concentrations were taken at mid-depth using a portable fluorometer, dye analyzer. Again, there is considerable scatter in measurements due to large-scale eddies in the receiving water and experimental error. Notice that the UDKHG predictions vary from 2 times too high to 2.5 times too low compared to data but on the whole it gave the correct trend and fell in the middle of the data. The CORMIX predictions were consistently low.

Figures 7.3 and 7.4 are plots of predicted versus measured dilutions at the SYOP line diffuser.[50] Figure 7.3 is for UM and 7.4 is for CORMIX2. Each point represents a different operating or ambient condition. The 45° line represents perfect agreement. Note that the UM predictions varied from three times too high to two times too low

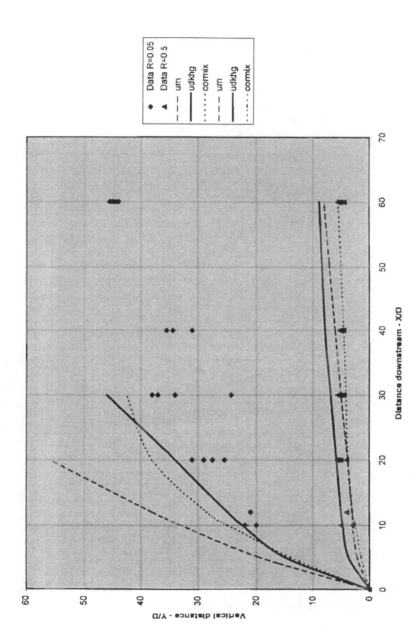

FIGURE 7.1 UDKHG, UM, and CORMIX trajectory predictions compared to laboratory data of Kannberg and Davis[13] for a multiple port line diffuser at two different currents.

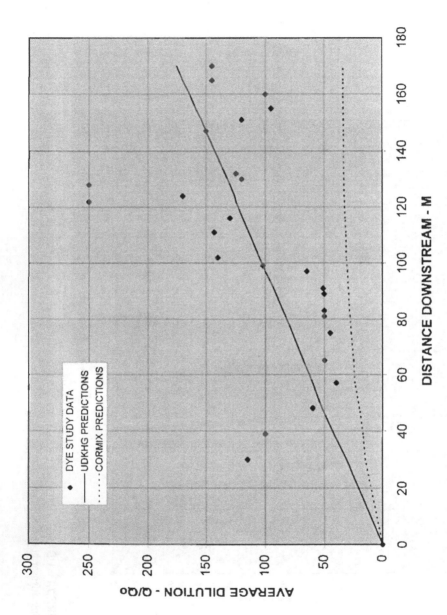

FIGURE 7.2   Average dilution as predicted by UDKHG and CORMIX versus distance as compared a dye study in the Cook Inlet[49] with depth = 4.6 m, current = 0.4 m/s, and discharge = 1.1 m³/s.

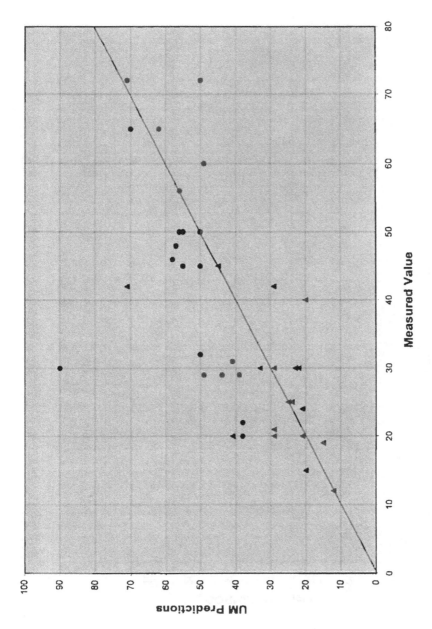

FIGURE 7.3 Predictions of UM compared to data at the SYOP outfall by Fergen et al.[50]

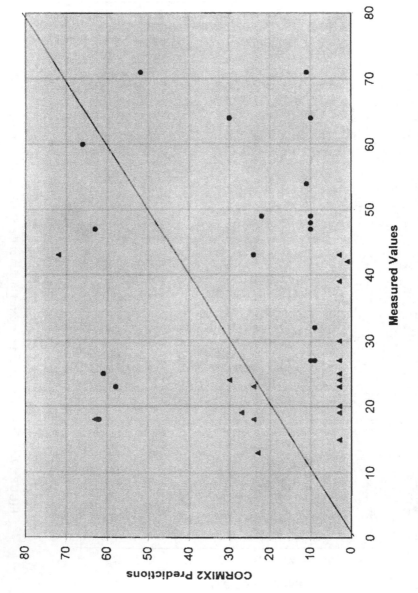

FIGURE 7.4 Predictions of CORMIX2 compared to data at the SYOP outfall by Fergen et al.[50]

but generally agreed with the data, while the CORMIX2 predictions had considerably more deviation and scatter.

Figure 7.5 is a plot of centerline surface dimensionless excess temperature as predicted by PDSG and CORMIX3 compared to the laboratory surface discharge study of Stefan et al.[51] Both PDSG and CORMIX3 agree fairly well in the near field, but PDSG is better in the far field. Finally, Figure 5.21 given earlier shows UDKHG predictions versus CORJET predictions for a multport diffuser discharged into a flowing stratified ambient. As can be seen, the two models give nearly the same values for both trajectory and concentration.

7.2   NUMERICAL MODELS

Just a comment on numerical models. By these I mean finite difference or finite element models that divide the ambient into a large number (but finite number) of nodes. Boundary and initial conditions are specified and the equations of motion are satisfied for each node as coupled to its neighboring nodes. These models have great potential in their ability to model transient conditions and variable geometry. As computer power and memory have increased, these models have become more of an option. They presently have two main problems. One is the inability to accurately model turbulence and the other is the difficulty in setting them up. Great advances have been made in the ability to model turbulence, but at present these models are no better and in some cases worse than the simpler integral models discussed above in predicting near field dilutions.[52,53] As a result, they have been mainly used to predict far field or transient behavior using the results of near field studies as input. A couple of examples are studies by Leedenertse and Liu[54] and Spalding.[55] There are several off-the-shelf commercial numerical models available. I won't pick any particular one. The trade magazines and displays at professional meetings will have representatives from the better ones. If you decide you can afford to get one, be sure and get one that is easy to set up, includes buoyancy, turbulence, three-dimensional and surface effects. It must also be able to predict both concentration and temperature. This will pretty much limit the ones you should consider.

7.3   RECOMMENDATIONS

Before I make any recommendations, let me state that these recommendations are based on my experience alone. They are my opinion and do not and are not intended to reflect any official policy! You would probably get a different set of recommendations if you talked to one of the developers of different models.

UM and WISP. Use the UM or WISP models for two-dimensional discharges of line diffusers where boundary effects are negligible. WISP is particularly handy for running a matrix of runs to study a variety of conditions. They are both very easy to set up and run. The three-dimensional approximation used by varying the effective port spacing is o.k. for vertical discharges where the jet–current interaction is included by the vertical angle, but the approximation underpredicts dilution for three-dimensional horizontal discharges where the jet–current interaction is ignored.

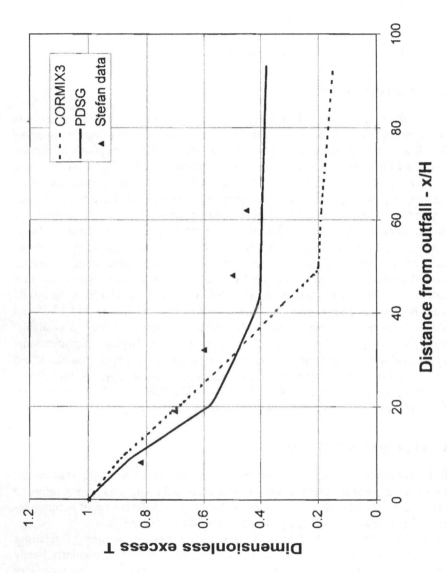

FIGURE 7.5 PDSG and CORMIX3 predictions versus experimental data of Stefan et al.[51]

RSB and RSBWIN. Use the RSB model to predict the initial dilution characteristics from diffusers in marine environments where the zone of initial dilution can be set by RSB. RSB is particularly useful in modeling diffusers with risers having multiple ports per riser with currents that vary from perpendicular to parallel to the outfall. It should not be used in rivers or other freshwater discharges or where the mixing zone is at a prescribed distance from the outfall.

UDKHG. Use UDKHG to predict plume characteristics from single or multiple port diffusers in two- or three-dimensional cases as long as boundary effects are negligible. The user must manually determine when the plume hits boundaries other than the surface. The graphical output is particularly useful in visualizing the plume and concentrations. The image method can take advantage of UDKHG's merging procedure to simulate shallow discharges of single port discharges. The ability to handle variable ambient stratification and vertical current in both UDKHG and UM is better than the approximations used in other models.

PDSG. Use the PDSG model for all surface discharges where boundary effects are negligible. The user must manually determine when the plume hits boundaries. The graphic output is very useful in visualizing the surface plume and determining when the plume interacts with boundaries.

CORMIX. The major strength of the CORMIX models is in the ability to handle complex cases and cases with boundary effects. I usually use one of the other models if they apply. They are easier to run and usually more accurate (unless CORJET is used). Of course, CORMIX only uses the CORJET model for the simpler cases and empirical equivalent slot models for the more complicated cases. If the other models don't apply or run into trouble, I use CORMIX.

ANL/UI. A lot more should be said about atmospheric discharges than I have in this book. I have really only discussed Gaussian plume modes briefly and the ANL/UI cooling tower plume model. Maybe in another book! The easiest way to determine plume characteristics from stacks and other environmental discharges without moisture effects is to use CORJET or the Gaussian plume models. If you are concerned about moisture, get the ANL/UI or SACTI model and the accompanying manuals from the Argonne National Laboratory. It seems to be the best at present.

References

1. Albertson, M. L., Dai, Y. B., Jensen, R. A., and Rouse, H., Diffusion of submerged jets, *ASCE Transactions*, Vol. 115, 639-697, 1970.
2. U.S. Environmental Protection Agency, Technical Support Document, EPA- 430/9-820--1, Sections 301(g) for freshwater and 301(h) for seawater of the Clean Water Act (PL 97-117), 1982.
3. U.S. Environmental Protection Agency, Technical Support Document for Water Quality-Based Toxics Control, USEPA, Office of Water, EPA/505/2-90-001, PB91-127415, Washington, D.C. 1991.
4. Muellenhoff, W. P., Schuldt, M. D., Baumgartner, D. J., Soldat, A. M., Davis, L. R., and Frick, W. E., Initial mixing characteristics of municipal ocean discharges, Procedures and Applications, EPA/600/3-85/073a, November 1985.
5. Teeter, A. M., and Baumgartner, D. J., Prediction of initial mixing for municipal ocean discharges, USEPA Corvallis Environmental Research Laboratory report CERL-043, May 1979.
6. Liseth, P., Mixing of merging jets from a manifold in stagnant receiving water of uniform density, University of California, Berkeley Hydraulic Engineering Laboratory Report HEL 231, November 1970.
7. Akar, P. J., and Jirka, G. H., CORMIX2: An expert system for hydrodynamic mixing zone analysis of conventional and toxic multiport diffuser discharges, U.S. Environmental Protection Agency Report, EPA/600/3-91/073, 1991.
8. Grace, R. A., *Marine Outfall Systems: Planning, Design, and Construction*, Prentice-Hall, Englewood Cliffs, N.J., 1978.
9. Colebrook, C. F., Turbulent flow in piles with particular reference to the transition region between smooth and rough pipe laws, *Journal of the Institute of Civil Engineers*, Vol. 11, 133, 1938-39.
10. Swamee, P. K., and Jain, A. K., Explicit equations for pipe-flow problems, *Journal of the Hydraulic Division*, HY5, 657, May 1976.
11. Baumgartner, D. J., Frick, W. E., and Roberts, P. J. W., Dilution models for effluent discharges, EPA/600/R-93/139, July 1993.
12. Brooks, N. H., Diffusion of sewage effluent in an open current, *Proceedings of the First Conference on Waste Disposal in the Marine Environment*, Ed. Pearson, E. A., Pergamon Press, New York, 569, 1960.
13. Kannberg, L. D., and Davis, L. R., An analysis of deep submerged multiple-port buoyant discharges, Transactions of ASME *Journal of Heat Transfer*, Vol. 98, Series C, No. 3, 367, August 1976.
14. Davis, L. R., Application of EPA computer models in buoyant plume prediction, *Heat Transfer in Convective Flows*, ASME Symposium Volume, HTD-Vol. 139, June 1989.
15. Dunn, W., Policastro, A. J., and Paddock, R. A., Surface thermal plumes: Evaluation of mathematical models for the near and complete field, Argonne National Laboratory Report ANL/EX-11, January 1975.
16. Hirst, E. A., Analysis of round turbulent buoyant jets discharged to flowing stratified ambients, Oak Ridge National Laboratory Report ORNL 4685, June 1971.
17. Hirst. E. A., Analysis of buoyant jets within the zone of flow establishment, Oak Ridge National Laboratory Report ORNL-TM-3470, August 1971.

18. Hoevekamp, T. B., *Buoyant Flow Simulation Programs with Interactive Graphics,* M.S. Thesis, Oregon State University, Corvallis, 1995.

19. Hsiao, E., *An Experimental/Analytical Investigation of Buoyant Jets in Shallow Water,* Ph.D. Thesis, Oregon State University, 1990.

20. Davis, L. R., and Hsiao, E., An experimental/analytical investigtion of buoyant jets in shallow water, *Proceedings of the First International Symposium on the Measurement and Modeling of Environmental Flows,* ASME Symposium Vol FED-Vol. 143, HTD-Vol 232, 217, November 1992.

21. Prych, E. A., A warm water effluent analyzed as a buoyant surface jet, Svergis Meterologiska Och Hydrologiska Institut, Serie Hydrologi. Nr 21, Stockholm, 1972.

22. Shirazi, M. A., and Davis, L. R., *Workbook of Thermal Plume Prediction, Vol. II: Surface Discharge,* EPA Report, EPA-R2-005a, May 1972b.

23. Davis, L. R., and Hoevekamp, T. B., A review of thermal plume simulation programs used in mixing zone analysis. *Proceedings of the Japan-U.S. Seminar on Thermal Engineering for Global Environmental Protection.* Invited paper No. A-9, Ed. Sengupta, S., San Francisco, CA, 1995.

24. Mullenhoff, W. P., Soldat, A. M., Davis, L. R., Baumgartner, M. D., Schuldt, M. D., and Frick, W. E., Initial mixing characteristics of municipal ocean discharges: Vol. II, Computer programs, EPA Report EPA-600/3/073b, November 1985.

25 Jirka, G. H., Doneker, R. L., and Hinton, S. W., User's manual for CORMIX: A hydrodynamic mixing zone model and decision support system for pollutant discharges into surface waters, USEPA, 1996. Available on the INTERNET at ftp://ftp.epa.gov/epa_ceam/wwwhtml/cormix.htm.

26. Roberts, P. J. W., Snyder, W. H., and Baumgartner, D. J., Ocean outfalls I: Submerged wastefield formation, *ASCE Journal of Hydraulic Engineering,* Vol. 115, No. 1, 1, January 1989.

27. Roberts, P. J. W., Snyder, W. H., and Baumgartner, D. J., Ocean outfalls II: Spacial evaluation of submerged wastefield, *ASCE Journal of Hydraulic Engineering.* Vol. 115, No. 1, 26, January 1989.

28. Roberts, P. J. W., Snyder, W. H., and Baumgartner, D. J., Ocean outfalls III: Effects of diffuser design on submerged wastefield, *ASCE Journal of Hydraulic Engineering,* Vol. 115, No. 1, 49, January 1989.

29. Jirka, G., and Fong, L. M., Vortex dynamics and bifurcation of buoyant jets in crossflow, *ASCE Journal of the Engineering Mechanics Division,* Vol. 107, No. EM3, 1981.

30. Macduff, R. B, and Davis, L. R., Multiple cell mechanical draft cooling tower model, *Environmental Effects of Atmospheric Heat/Moisture Releases,* Proceedings of the 2nd AIAA/ASME Thermophysics and Heat Transfer Symposium, Palo Alto, CA, May 1978.

31. Winiariski, L., and Frick, W., Methods of improving plume models. Cooling tower environment, *Proceedings of Power Plant Siting Commission.* Maryland Department of Natural Resources. PPSP-CPCTP-22. WRRC Special Report No. 9, 1978.

32. Carhart, R. A., and Policastro, A. J., A second-generation model for cooling tower plume rise and dispersion — I. Single sources, *Atmospheric Environment,* Vol. 25A, No. 8, 1559, 1991.

33. Hanna, S. R., Predicted and observed cooling tower plume rise and visible plume length at the John E. Amos power plant, *Atmospheric Environment,* Vol. 10, 1043, 1975.

34. Orville, H. D., Hirsch, J. H., and May, L. E., Application of a cloud model to cooling tower plumes and clouds, *Journal of Applied Meteriology,* Vol. 19, 1260, 1980.

35. Slawson, P., and Wigley, P. M., The effects of atmospheric specific conditions on the length of visible cooling tower plumes, *Atmospheric Environment*, Vol. 9, 437, 1975.
36. Dunn, W. E., Coke, L., and Policastro, A. J., User's manual: A computerized methodology for predicting seasonal/annual impacts of visible plumes, drift, fogging, icing, and shadowing from single and multiple sources, Research Project Report 906-1, Electric Power Research Institute, September 1987.
37. Carhart, R. A., Policastro, A. J., and Dunn, W. E., An improved method for predicting seasonal and annual shadowing from cooling tower plumes, *Atmospheric Environment*, Vol. 26a, No. 15, 2845, 1992.
38. Policastro, A. J., Dunn, W. E., and Carhart, R. A., A model for seasonal and annual cooling tower impacts, *Atmospheric Environment*, Vol. 28, No. 3, 379, 1994.
39. Turner, D. B., *Workbook of Atmospheric Dispersion Estimates*, PHS Publication No. 999-26, U.S. EPA, 1969.
40. Briggs, G. A., *Plume Rise*, U.S. Atomic Energy Commission, NTIS No. TID-25075, 1969.
41. Viollet, M., P-L, Étude de jets dans des courants traversiers et dans des milieux stratifiés, These de Docteut-Ingénieur, Université P. et M. Curie, Paris, 1977.
42. Bremer, P., Berict über die meteorologische meserie am standort Lünen vom 27.11.1972 bis 1.12.1972 zur teilüberprüfung und tereinerung des numerischen models SAUNA, Arbeitsgruppe über die meteorologischen Auswirkunger der Kühltürme, *Dienst für Luftreinhaltung der Schweizerischen Meteorologischen Zentralanstalt*, Payern, 1973.
43. Kannberg, L. D, and Onishi, Y., Plumes from one and two cooling towers, *Environmental Effects of Atmospheric Heat/Moisture Releases. Cooling Towers, Cooling Ponds, and Area Sources*, Ed. Torrance, K., and Watts, R., ASME, New York, 1978.
44. Meyer, J. H., and Jenkins, W. R., Chalk Point surface weather and ambient atmospheric profile data, second intensive test period, 14-24 June, 1976, Applied Physics Lab., Johns Hopkins Univ., PPPSP-CPCTP-4REV, 1975.
45. Slawson, P. R., and Coleman, J. H., Natural draft cooling tower plume behavior at Paradise steam plant. Part II, Tennessee Valley Authority, Division of Environmental Planning, TVA/EP-78-01, February 1978.
46. Moore, R. E., The KUMULUS model for plume drift deposition calculations for Indian Point Unit No. 2, Environmental System Corporation, Knoxville, TN, 1977.
47. Wu, F. H., and Koh, R. C., Mathematical model for multiple cooling tower plumes, W. M. Keck Laboratory of Hydraulics and Water Resources, Report KH-R-37, California Institute of Technology, Pasadena, CA, 1977.
48. Dunn, W. E., Gavin, P., and Boughton, B., Studies on mathematical models for characterizing plume and drift behavior from cooling towers, Vol. 3. Mathematical model for single-source cooling tower drift dispersion, Electric Power Research Institute Report, CS-1683, Palo Alto, CA, 1981.
49. Lindquist, R., Personal communication, CH2M HILL, Corvallis, OR, 1988.
50. Fergen, R. E., Huang, H., and Proni, J. R., *Proceedings of the WEFTEC Symposium*, Chicago, 1994.
51. Stefan, H., Hayakawa, N., and Shiebe, R. R., Surface discharge of heated water, EPA Water Pollution Conntrol Series Report, 16130, FSU, 12/71, 1971.
52. Dunn, W., Policastro, A. J., and Paddock, R. A., Surface thermal plumes: Evaluation of mathematical models for the near and complete field, Argonne National Lab, report ANL/WR-75-3 Part 1, May 1975, Part 2, August 1975.

53. Policastro, A. J., and Dunn, W. E., Numerical modeling of surface thermal plumes, ICHMT, International Advanced Course on Heat Disposal from Power Generation, Dubrovnik, Yugoslovia, 1976.

54. Leedenertse, J. J., and Liu, S., A three-dimensional model for estuaries and coastal seas: Vol. I, Principles of computation, The Rand Corporation Report R-1417-OWRR, 1973.

55. Spalding, D. B., THRIBLE — Transfer of heat in rivers, bays, lakes and estuaries. Department of Mechanical Engineering, Imperial College, Heat Transfer Section report HTS/75/4, London, 1975.

APPENDIX I
Definition of Terms
Used in CORMIX

A0	Nozzle area
AREG	Regulatory mixing zone area
B	Gaussion (37%) half width — CORJET
B0	Initial equivalent slot half width
BETA	Discharge angle relative to diffuser line
BETYPE	Nozzle arrangement
BH	Plume horizontal size — definition varies from module to module
BS	Bounded ambient effective width
BV	Plume vertical size — definition varies from module to module
C0	Initial concentration
Cc	Centerline concentration
CCC	Criterion continuous concentration for edge of mixing zone
CMC	Criterion maximum concentration for edge of zone of initial dilution
CSTD	Regulatory mixing zone concentration standard
CUNITS	Units of concentration
D0	Nozzle diameter
DISMAX	Maximum distance for simulation
DIST	Distance along plume trajectory
DISTB	Distance from edge of simulated bank or shore to the discharge location — for diffusers, it's to the center of the diffuser.
DITYPE	Diffuser type
DRHO	Initial density difference
dSALc	Centerline excess salinity (CORJET)
dTc	Centerline excess temperature (CORJET)
F	Darcy–Weisbach friction factor
Flc	Local Froude number
FLOCLS	Flow class as determined from length scales
FR0	Discharge densimetric Froude number (port for CORJET and slot for CORMIX equivalent slot)
FRD0	Port densimetric Froude number = $U0/(GP0 \cdot D0)^{1/2}$
GAMMA	Horizontal discharge direction relative to X-coordinate with CCW positive
GAMMAE	Gamma angle at end of flow establishment region
GP0	Discharge buoyancy = $g(\rho_a - \rho_o)/\rho_a$
Gpc	Local centerline buoyancy term, $g' = g(\rho_a - \rho)/\rho_a$
H0	Discharge channel depth in CORMIX3
HA	Simulated average depth (m)
HD	Actual discharge depth (m)
HINT	Location of pycnocline above bottom (m)
ICHREG	Channel regularity flag

IPOLL	Pollutant type
j0	Kinetic buoyancy flux per unit length, $g_o' q_o$
J0	Kinetic buoyancy flux, $g_o' U_o$
KD	Decay coefficient per hour
la	Crossflow/stratification length scale U_a/N
Lb	Plume/crossflow length scale, J_o/U_a^3
Lbp	Plume/stratification length scale, $J_o/N^{3/4}$
lb'	Slot plume/stratification length scale, $j_o^{1/3}/N$
LD	Diffuser length (m)
LE	Length of flow development zone in CORJET
LEV	Level number (CORJET)
lm	Slot jet/crossflow length scale, $U_o q_o/U_a^2$
lM	Slot jet/plume transition length scale, $U_o q_o/j_o^{2/3}$
Lm	Jet/crossflow length scale, $(U_o Q_o)^{1/2}/U_a$
LM	Jet/plume transition length scale, $(U_o Q_o)^{3/4}/J_o^{1/2}$
Lmp	Jet/stratification length scale, $(U_o Q_o)^{1/4}/N^{1/2}$
lm'	Slot jet/stratification length scale, $(U_o q_o)^{1/3}/N$
lq	Flow length scale for slot (RSB)
LQ	Flow length scale for jet
m0	Momentum flux — slot
M0	Momentum flux — jet or channel
NFR	Near field region
NOPEN	Number of openings
NPRINT	Number of printout values
NSTD	Regulatory mixing zone flag, true if 1.0, false if 0
NTOX	Toxic effluent flag, true if 1.0, false if 0
q0	Diffuser discharge per unit length, Q_o/L
Q_o	Total discharge rate (m³/s)
R	Velocity ratio, U_o/U_a
REGMZ	Value of regulatory mixing zone
REGSPC	Regulatory mixing zone standard
RHOO	Discharge density, ρ_o
RHOAB	Density of ambient at bottom
RHOAH0	Density of ambient at pycnocline
RHOAS	Density of ambient at surface
ROI	Region of interest
RMZ	Regulatory mixing zone
S	Dilution $Q/Q0$
Save	Average bulk dilution
Sc	Centerline dilution
SIGMA	Diffuser orientation relative to X-axis, CCW positive, 0 parallel, 90 or 180 perpendicular
SIGMA0	Sigma angle of nozzle (CORJET)
SIGMAE	Sigma angle at end of zone of flow establishment (CORJET)
SPAC	Port spacing (m)
STRCND	Stratification condition, A, B, or C
TDZ	Toxic dilution (mixing) zone
THETA	Vertical discharge angle, 0 horizontal, 90 vertical
THETA0	Initial jet vertical discharge angle
THETAE	Vertical discharge angle at end of flow establishment, CORJET
U0	Discharge velocity (m/s)

UA	Ambient velocity (m/s)
UW	Wind speed (m/s)
WREG	Width of regulatory mixing zone
X	Horizontal coordinate downstream in direction of current
XE	Distance in X direction to end of zone of flow establishment, CORJET
XINT	Region of interest (m)
XMAX	Maximum region of simulation (m)
XREG	Extent of regulatory mixing zone (m)
Y	Horizontal coordinate across current
YB1	Distance in Y direction from nearest bank to the first diffuser port (m)
YB2	Distance in Y direction from nearest bank to the furthest diffuser port (m)
YE	Distance in Y direction to end of zone of flow establishment, CORJET
Z	Vertical coordinate above discharge (distance up)
ZA	Ambient table entry level (CORJET)
ZE	Distance in Z direction to end of zone of flow establishment, CORJET
ZL	Lower Z
ZMAX	Maximum Z of interest (CORJET)
ZMIN	Minimum Z of interest (CORJET)
ZU	Upper Z

Appendix II

CORMIX Prediction File

```
CORMIX2 PREDICTION FILE:
22222222222222222222222222222222222222222222222222222222222222222222222222
                      CORNELL MIXING ZONE EXPERT SYSTEM
Subsystem CORMIX2:                                        Subsystem version:
 Submerged Multiport Diffuser Discharges    CORMIX_v.3.20_____September_1996
----------------------------------------------------------------------------


----------------------------------------------------------------------------


CASE DESCRIPTION
 Site name/label:       Example 5.3
 Design case:           300-4^inch^ports
 FILE NAME:             cormix\sim\3004alt .cx2
 Time of Fortran run:   03/05/98--09:48:08

ENVIRONMENT PARAMETERS (metric units)
 Unbounded section
 HA    =      95.00  HD    =      96.00
 UA    =       .044  F     =       .002  USTAR = .6460E-03
 UW    =       .000  UWSTAR= .0000E+00
 Density stratified environment
 STRCND=  A          RHOAM = 1022.0000
 RHOAS = 1021.0000  RHOAB = 1023.0000  RHOAH0= 1022.9790  E     = .2000E-03

DIFFUSER DISCHARGE PARAMETERS (metric units)
 Diffuser type:      DITYPE= alternating_perpendicular
 BANK  =  LEFT       DISTB =    3150.00  YB1   =   3000.00  YB2  =   3300.00
 LD    =     300.00  NOPEN =  148        SPAC  =      2.04
 D0    =       .100  A0    =       .008  H0    =      1.00
 Nozzle/port arrangement:   alternating_without_fanning
 GAMMA =      90.00  THETA =      90.00  SIGMA =       .00  BETA =     90.00
 U0    =       .516  Q0    =       .600  =             .6000E+00
 RHO0  =  999.0000  DRHO0 = .2398E+02  GP0   = .2299E+00
 C0    = .1000E+04  CUNITS= percent
 IPOLL =  1          KS    = .0000E+00  KD    = .0000E+00


FLUX VARIABLES - PER UNIT DIFFUSER LENGTH (metric units)
 q0    = .2000E-02  m0    = .1032E-02  j0    = .4598E-03  SIGNJ0=     1.0
 Associated 2-d length scales (meters)
 lQ=B  =       .004  lM    =       .17  lm    =       .53
 lmp   =      1.73  lbp   =      5.46  la    =      3.11


FLUX VARIABLES - ENTIRE DIFFUSER (metric units)
 Q0    = .6000E+00  M0    = .3097E+00  J0    = .1379E+00
 Associated 3-d length scales (meters)
 LQ    =      1.08  LM    =       1.12  Lm    =     12.65  Lb   =   1619.13
                                        Lmp   =      6.27  Lbp  =     14.86


NON-DIMENSIONAL PARAMETERS
 FR0   =      17.29  FRD0  =       3.40  R     =     11.73
 (slot)             (port/nozzle)
```

```
FLOW CLASSIFICATION
22222222222222222222222222222222222222222222
2  Flow class (CORMIX2)      =    MS5    2
2  Applicable layer depth HS =    96.00  2
22222222222222222222222222222222222222222222

MIXING ZONE / TOXIC DILUTION / REGION OF INTEREST PARAMETERS
CO    = .1000E+04  CUNITS=  percent
NTOX  = 0
NSTD  = 1          CSTD  = .1000E+01
REGMZ = 1
REGSPC= 1          XREG  =    150.00  WREG =        .00  AREG =        .00
XINT  =  9500.00  XMAX  =   9500.00

X-Y-Z COORDINATE SYSTEM:
     ORIGIN is located at the bottom and the diffuser mid-point:
        3150.00 m  from the LEFT  bank/shore.
     X-axis points downstream, Y-axis points to left, Z-axis points upward.
NSTEP = 20 display intervals per module
-----------------------------------------------------------------------------


-----------------------------------------------------------------------------


BEGIN MOD101: DISCHARGE MODULE (SINGLE PORT AT DIFFUSER CENTER)

     Initial conditions for individual jet/plume:
     Average spacing between jet/plumes:    2.04 m
        X        Y        Z        S        C        BV        BH
       .00      .00     1.00     1.0   .100E+04    .05       .05

END OF MOD101: DISCHARGE MODULE (SINGLE PORT AT DIFFUSER CENTER)
-----------------------------------------------------------------------------


-----------------------------------------------------------------------------


BEGIN CORJET (MOD110): JET/PLUME NEAR-FIELD MIXING REGION

Plume-like motion in linear stratification with strong crossflow.

Zone of flow establishment:              THETAE=    80.64  SIGMAE=      .00
LE    =       .00  XE    =      .00  YE   =       .00  ZE    =     1.00

Profile definitions:
     BV = Gaussian 1/e (37%) half-width, in vertical plane normal to trajectory
     BH = before merging: Gaussian 1/e (37%) half-width in horizontal plane
                          normal to trajectory
          after merging:  top-hat half-width in horizontal plane
                          parallel to diffuser line
     S  = hydrodynamic centerline dilution
     C  = centerline concentration (includes reaction effects, if any)

        X        Y        Z        S        C        BV        BH
Individual jet/plumes before merging:
       .00      .00     1.00     1.0   .100E+04    .05       .05
```

```
        .37      .00    2.34     9.0  .112E+03    .24      .24
        .92      .00    3.61    25.1  .398E+02    .46      .46
       1.59      .00    4.83    47.4  .211E+02    .66      .66
       2.47      .00    5.89    74.7  .134E+02    .86      .86
       3.70      .00    6.53    98.8  .101E+02   1.01     1.01
Merging of individual jet/plumes to form plane jet/plume:
       5.03      .00    6.94   142.1  .704E+01   1.39   151.39
       6.38      .00    7.27   153.5  .651E+01   1.51   151.51
       7.74      .00    7.58   164.8  .607E+01   1.63   151.63
       9.09      .00    7.88   175.8  .569E+01   1.76   151.76
      10.45      .00    8.17   186.5  .536E+01   1.89   151.89
      11.82      .00    8.44   197.1  .507E+01   2.02   152.02
      13.18      .00    8.70   207.3  .482E+01   2.15   152.15
      14.55      .00    8.95   217.2  .460E+01   2.27   152.27
      15.92      .00    9.18   226.7  .441E+01   2.40   152.40
      17.29      .00    9.40   235.9  .424E+01   2.53   152.53
      18.67      .00    9.61   244.6  .409E+01   2.65   152.65
      20.05      .00    9.79   252.8  .396E+01   2.77   152.77
      21.43      .00    9.96   260.4  .384E+01   2.89   152.89
      22.81      .00   10.11   267.0  .375E+01   2.99   152.99
      24.19      .00   10.24   272.2  .367E+01   3.07   153.07
Terminal level in stratified ambient has been reached.
Cumulative travel time =      225. sec

END OF CORJET (MOD110): JET/PLUME NEAR-FIELD MIXING REGION
----------------------------------------------------------------------------

----------------------------------------------------------------------------

BEGIN MOD235: LAYER/BOUNDARY/TERMINAL LAYER APPROACH

Control volume inflow:
     X         Y        Z        S        C        BV       BH
   24.19      .00   10.24   272.2  .367E+01   3.07   153.07

Profile definitions:
  BV = top-hat thickness, measured vertically
  BH = top-hat half-width, measured horizontally in y-direction
  ZU = upper plume boundary (Z-coordinate)
  ZL = lower plume boundary (Z-coordinate)
  S  = hydrodynamic average (bulk) dilution
  C  = average (bulk) concentration (includes reaction effects, if any)

     X         Y        Z        S        C        BV       BH      ZU      ZL
   21.12      .00   10.24   272.2  .367E+01    .00      .00   10.24   10.24
   22.96      .00   10.24   272.2  .367E+01  25.07    69.83   22.77     .00
   24.81      .00   10.24   291.8  .343E+01  29.44    98.76   24.96     .00
   26.65      .00   10.24   507.5  .197E+01  31.88   120.95   26.18     .00
   28.49      .00   10.24   699.9  .143E+01  33.20   139.66   26.84     .00
   30.34      .00   10.24   770.0  .130E+01  33.62   156.15   27.05     .00
Cumulative travel time =      365. sec

END OF MOD235: LAYER/BOUNDARY/TERMINAL LAYER APPROACH
----------------------------------------------------------------------------
```

```
** End of NEAR-FIELD REGION (NFR) **
-------------------------------------------------------------------------

BEGIN MOD242: BUOYANT TERMINAL LAYER SPREADING

Profile definitions:
   BV = top-hat thickness, measured vertically
   BH = top-hat half-width, measured horizontally in y-direction
   ZU = upper plume boundary (Z-coordinate)
   ZL = lower plume boundary (Z-coordinate)
   S  = hydrodynamic average (bulk) dilution
   C  = average (bulk) concentration (includes reaction effects, if any)

Plume Stage 1 (not bank attached):
      X        Y       Z       S        C        BV       BH       ZU       ZL
   30.34      .00    10.24   770.0   .130E+01   33.62   156.15   27.05     .00
** WATER QUALITY STANDARD OR CCC HAS BEEN FOUND **
The pollutant concentration in the plume falls below water quality standard
   or CCC value of   .100E+01 in the current prediction interval.
This is the spatial extent of concentrations exceeding the water quality
   standard or CCC value.
** REGULATORY MIXING ZONE BOUNDARY **
In this prediction interval the plume distance meets or exceeds
   the regulatory value =   150.00 m.
This is the extent of the REGULATORY MIXING ZONE.
  167.33      .00    10.24   1113.9   .898E+00   11.20   677.89   15.84    4.64
  304.32      .00    10.24   1222.0   .818E+00    8.55   974.61   14.51    5.96
  441.31      .00    10.24   1291.1   .775E+00    7.29  1207.05   13.88    6.59
  578.30      .00    10.24   1342.4   .745E+00    6.52  1403.81   13.50    6.98
  715.29      .00    10.24   1383.5   .723E+00    5.98  1576.88   13.23    7.25
  852.28      .00    10.24   1417.9   .705E+00    5.58  1732.67   13.03    7.45
  989.27      .00    10.24   1447.4   .691E+00    5.26  1875.16   12.87    7.61
 1126.26      .00    10.24   1473.3   .679E+00    5.00  2006.98   12.74    7.74
 1263.25      .00    10.24   1496.4   .668E+00    4.79  2130.00   12.63    7.84
 1400.24      .00    10.24   1517.2   .659E+00    4.61  2245.60   12.54    7.93
 1537.23      .00    10.24   1536.2   .651E+00    4.45  2354.86   12.46    8.01
 1674.22      .00    10.24   1553.6   .644E+00    4.31  2458.60   12.39    8.08
 1811.21      .00    10.24   1569.8   .637E+00    4.18  2557.49   12.33    8.15
 1948.20      .00    10.24   1584.8   .631E+00    4.07  2652.07   12.27    8.20
 2085.19      .00    10.24   1598.8   .625E+00    3.97  2742.80   12.22    8.25
 2222.18      .00    10.24   1612.0   .620E+00    3.88  2830.06   12.18    8.30
 2359.17      .00    10.24   1624.4   .616E+00    3.80  2914.16   12.14    8.34
 2496.16      .00    10.24   1636.1   .611E+00    3.72  2995.40   12.10    8.38
 2633.15      .00    10.24   1647.3   .607E+00    3.65  3073.99   12.06    8.41
 2770.14      .00    10.24   1657.9   .603E+00    3.59  3150.16   12.03    8.44
Cumulative travel time =      62633. sec

-------------------------------------------------------------------------

Plume is ATTACHED to LEFT  bank/shore.
   Plume width is now determined from LEFT  bank/shore.

Plume Stage 2 (bank attached):
```

X	Y	Z	S	C	BV	BH	ZU	ZL
2770.14	3150.00	10.24	1657.9	.603E+00	3.59	6300.00	12.03	8.44
3106.63	3150.00	10.24	1667.7	.600E+00	3.53	6443.67	12.00	8.47
3443.12	3150.00	10.24	1677.0	.596E+00	3.47	6583.02	11.97	8.50
3779.62	3150.00	10.24	1685.9	.593E+00	3.42	6718.35	11.95	8.53
4116.11	3150.00	10.24	1694.5	.590E+00	3.37	6849.96	11.92	8.55
4452.60	3150.00	10.24	1702.7	.587E+00	3.33	6978.10	11.90	8.57
4789.10	3150.00	10.24	1710.7	.585E+00	3.28	7103.00	11.88	8.60
5125.59	3150.00	10.24	1718.3	.582E+00	3.24	7224.85	11.86	8.62
5462.08	3150.00	10.24	1725.7	.579E+00	3.20	7343.85	11.84	8.64
5798.58	3150.00	10.24	1732.8	.577E+00	3.17	7460.16	11.82	8.65
6135.07	3150.00	10.24	1739.7	.575E+00	3.13	7573.94	11.80	8.67
6471.56	3150.00	10.24	1746.4	.573E+00	3.10	7685.31	11.79	8.69
6808.06	3150.00	10.24	1752.9	.570E+00	3.07	7794.41	11.77	8.70
7144.55	3150.00	10.24	1759.2	.568E+00	3.04	7901.36	11.76	8.72
7481.04	3150.00	10.24	1765.3	.566E+00	3.01	8006.26	11.74	8.73
7817.54	3150.00	10.24	1771.2	.565E+00	2.98	8109.21	11.73	8.75
8154.03	3150.00	10.24	1777.0	.563E+00	2.95	8210.30	11.71	8.76
8490.52	3150.00	10.24	1782.7	.561E+00	2.92	8309.62	11.70	8.78
8827.01	3150.00	10.24	1788.2	.559E+00	2.90	8407.25	11.69	8.79
9163.51	3150.00	10.24	1793.6	.558E+00	2.88	8503.25	11.68	8.80
9500.00	3150.00	10.24	1798.9	.556E+00	2.85	8597.71	11.66	8.81

Cumulative travel time =   215584. sec

Simulation limit based on maximum specified distance =   9500.00 m.
   This is the REGION OF INTEREST limitation.

END OF MOD242: BUOYANT TERMINAL LAYER SPREADING

CORMIX2: Submerged Multiport Diffuser Discharges   End of Prediction File
2222222222222222222222222222222222222222222222222222222222222222222222222222222

Index

Milton Keynes UK
Ingram Content Group UK Ltd.
UKHW040052071024
449327UK00019B/517